Speed Statistics: A Formula Handbook

N.B. Singh

DEDICATION

To Nature,

I dedicate this book to you, the source of all life. You are my inspiration, my teacher, and my friend.

Thank you for teaching me about the beauty of the world around me. Thank you for showing me the power of the natural world. Thank you for giving me a sense of peace and tranquillity.

I promise to do my part to protect you and your many wonders. I will teach my children about the importance of conservation and sustainability. I will work to make the world a better place for all living things.

Thank you for everything, Nature.

With love,

N.B Singh

Contents

Welcome to "Speed Statistics: A Formula Handbook." This handbook is crafted with the intention of providing a quick and comprehensive reference for individuals engaged in the field of statistics, ranging from students to seasoned professionals.

Objectives of the Handbook

The main objectives of this handbook are to offer a concise compilation of essential formulas and equations related to speed statistics and to facilitate a deeper understanding of statistical concepts. Whether you are a learner or a practitioner, we hope this handbook becomes a valuable resource in your statistical endeavors.

Organization of the Handbook

The handbook is organized systematically into various sections, each dedicated to a specific aspect of speed statistics. From measures of central tendency to advanced statistical methods, you will find a wealth of formulas presented in a clear and organized manner. Mathematical expressions are used extensively to ensure precision and clarity.

How to Use This Handbook

Navigating through the handbook is designed to be intuitive. Each section focuses on a specific topic, allowing for quick access to the information you need. Examples and applications are included to illustrate the practical use of the formulas, aiding in real-world applications.

Chapter 1

Introduction

1.1 Overview

In this section, we provide a quick overview of key concepts using mathematical formulas and equations.

$$\text{Mean:} \quad \bar{x} = \frac{1}{n} \sum_{i=1}^{n} x_i$$

$$\text{Variance:} \quad \sigma^2 = \frac{1}{n} \sum_{i=1}^{n} (x_i - \bar{x})^2$$

$$\text{Normal Distribution:} \quad f(x|\mu, \sigma^2) = \frac{1}{\sqrt{2\pi\sigma^2}} e^{-\frac{(x-\mu)^2}{2\sigma^2}}$$

$$\text{Hypothesis Testing:} \quad p - value = P(\text{observe the data}|\text{null hypothesis is true})$$

$$\text{Linear Regression:} \quad y = mx + b$$

$$\text{Bayes' Theorem:} \quad P(A|B) = \frac{P(B|A)P(A)}{P(B)}$$

This concise overview covers essential statistical concepts, providing a foundation for the detailed discussions in the subsequent chapters.

1.2   Importance of Speed Statistics

Speed statistics play a crucial role in understanding motion and making informed decisions. Here are key formulas to remember:

$$\text{Speed:} \quad \text{Speed} = \frac{\text{Distance}}{\text{Time}}$$

$$\text{Velocity:} \quad \text{Velocity} = \frac{\text{Displacement}}{\text{Time}}$$

$$\text{Acceleration:} \quad \text{Acceleration} = \frac{\text{Change in Velocity}}{\text{Time}}$$

$$\text{Kinematic Equations:} \quad v_f = v_i + at, \quad s$$
$$= v_i t + \frac{1}{2}at^2, \quad v_f^2$$
$$= v_i^2 + 2as$$

$$\text{Average Speed:} \quad \text{Average Speed} = \frac{\text{Total Distance}}{\text{Total Time}}$$

$$\text{Relative Velocity:} \quad v_{\text{rel}} = v_{\text{A}} - v_{\text{B}}$$

These formulas provide quick insights into motion and are essential for solving problems in various fields.

1.3   Scope and Objectives

The scope and objectives of this work are outlined with key mathematical formulations:

$$\text{Scope:} \quad \text{Scope}$$
$$= \{x$$
$$\mid x$$
$$\in \text{Domain}\}$$

$$\text{Objectives:} \quad \text{Maximize } f(x) \quad \text{subject to} \quad g_i(x)$$
$$\leq 0, i$$
$$= 1, 2, \ldots, m$$

$$\text{Constraint Equations:} \quad g_i(x) = 0 \quad \text{for} \quad i = 1, 2, \ldots, m$$

$$\text{Optimization Problem:} \quad \text{Minimize } f(x) \quad \text{subject to} \quad h_j(x) = 0, j = 1, 2, \ldots, r$$

$$\text{Decision Variables:} \quad x_1, x_2, \ldots, x_n$$

$$\text{Objective Function:} \quad f(x) = c^T x$$

These mathematical representations succinctly define the scope and objectives of the work, providing a foundation for the upcoming discussions.

1.4 Notation and Terminology

This section introduces key notations and terminology using mathematical expressions:

$$\text{Set Notation:} \quad A = \{a, b, c, \ldots\}$$

$$\text{Function Notation:} \quad f : X \to Y, \quad f(x) = y$$

$$\text{Summation Notation:} \quad \sum_{i=1}^{n} x_i$$

$$\text{Product Notation:} \quad \prod_{i=1}^{n} x_i$$

$$\text{Derivative Notation:} \quad f'(x), \quad \frac{dy}{dx}$$

$$\text{Integral Notation:} \quad \int_a^b f(x)\, dx$$

$$\text{Vector Notation:} \quad \mathbf{v} = [v_1, v_2, \ldots, v_n]$$

These notations and terminologies are essential for clear communication and understanding of mathematical concepts.

1.5   Background

The background section provides a quick overview of foundational concepts using mathematical expressions:

$$\text{Pythagorean Theorem:} \quad c^2 = a^2 + b^2$$

$$\text{Euler's Formula:} \quad e^{ix} = \cos(x) + i\sin(x)$$

$$\text{Quadratic Formula:} \quad x = \frac{-b \pm \sqrt{b^2 - 4ac}}{2a}$$

$$\text{Binomial Theorem:} \quad (a+b)^n = \sum_{k=0}^{n} \binom{n}{k} a^{n-k} b^k$$

$$\text{Taylor Series:} \quad f(x) = f(a) + f'(a)(x-a) + \frac{f''(a)}{2!}(x-a)^2 + \dots$$

$$\text{Bayes' Theorem:} \quad P(A|B) = \frac{P(B|A)P(A)}{P(B)}$$

These foundational mathematical concepts form the background for understanding more advanced topics.

1.6   Organization of the Book

The book is structured to facilitate efficient learning and understanding, incorporating key mathematical concepts:

$$\text{Chapter Sequencing:} \quad \text{Chapter}_n = \text{Chapter}_{n-1} + 1$$

$$\text{Section Dependencies:} \quad \text{Section}_i \rightarrow \text{Section}_{i+1}$$

$$\text{Key Themes:} \quad \text{Theme}_1, \text{Theme}_2, \dots, \text{Theme}_n$$

$$\text{Cross-References:} \quad \text{See Chapter}_i, \text{Section}_j$$

$$\text{Equation Numbering:} \quad E_{\text{Chapter}_n.\text{Section}_i.\text{Equation}_j}$$

$$\text{Quick Reference Index:} \quad \text{Index}_i = \frac{\text{Page}_i}{\text{Total Pages}}$$

This organizational structure aims to enhance accessibility and ease of navigation for effective learning.

Chapter 2

Descriptive Statistics

2.1   Measures of Central Tendency

This section introduces key measures summarizing central tendencies with memorable
mathematical expressions:

$$\text{Mean:} \quad \bar{x} = \frac{1}{n} \sum_{i=1}^{n} x_i$$

$$\text{Weighted Mean:} \quad \bar{x}_w = \frac{\sum_{i=1}^{n} w_i x_i}{\sum_{i=1}^{n} w_i}$$

$$\text{Median:} \quad \text{Median} = \begin{cases} x_{(n+1)/2} & \text{for odd } n \\ \frac{x_{n/2} + x_{n/2+1}}{2} & \text{for even } n \end{cases}$$

$$\text{Mode:} \quad \text{Mode} = \text{Value with highest frequency}$$

$$\text{Geometric Mean:} \quad GM = \left(\prod_{i=1}^{n} x_i \right)^{1/n}$$

$$\text{Harmonic Mean:} \quad HM = \frac{n}{\sum_{i=1}^{n} \frac{1}{x_i}}$$

These measures provide insights into the central tendencies of a dataset in
various ways.

2.2   Measures of Dispersion

This section introduces key measures summarizing dispersion with memorable mathematical expressions:

$$\text{Range:} \quad \text{Range} = \max(X) - \min(X)$$

$$\text{Interquartile Range (IQR):} \quad \text{IQR} = Q_3 - Q_1$$

$$\text{Variance:} \quad \sigma^2 = \frac{1}{n} \sum_{i=1}^{n} (x_i - \bar{x})^2$$

$$\text{Standard Deviation:} \quad \sigma = \sqrt{\sigma^2}$$

$$\text{Coefficient of Variation (CV):} \quad CV = \left(\frac{\sigma}{\bar{x}}\right) \times 100\%$$

$$\text{Mean Absolute Deviation (MAD):} \quad MAD = \frac{1}{n} \sum_{i=1}^{n} |x_i - \bar{x}|$$

These measures offer insights into the spread or dispersion of values within a dataset.

2.3   Percentiles and Quartiles

This section introduces key concepts related to percentiles and quartiles with memorable mathematical expressions:

$$\text{Percentile:} \quad P_k = \frac{k}{100}(n + 1)$$

$$\text{Lower Quartile (Q1):} \quad Q_1 = \frac{1}{4}(n + 1)$$

$$\text{Median (Q2):} \quad Q_2 = \frac{1}{2}(n + 1)$$

$$\text{Upper Quartile (Q3):} \quad Q_3 = \frac{3}{4}(n + 1)$$

$$\text{Interquartile Range (IQR):} \quad \text{IQR} = Q_3 - Q_1$$

$$\text{Five-Number Summary:} \quad \{\min(X), Q_1, Q_2, Q_3, \max(X)\}$$

These concepts provide a quick and effective way to understand the distribution of data in a dataset.

2.4   Skewness and Kurtosis

This section introduces key concepts related to skewness and kurtosis with memorable mathematical expressions:

$$\text{Skewness:} \quad \text{Skewness} = \frac{\sum_{i=1}^{n}(x_i - \bar{x})^3}{n \cdot \sigma^3}$$

$$\text{Pearson's First Coefficient of Skewness:} \quad \text{Skewness} = 3\left(\frac{\bar{x} - \mu}{\sigma}\right)$$

$$\text{Kurtosis:} \quad \text{Kurtosis} = \frac{\sum_{i=1}^{n}(x_i - \bar{x})^4}{n \cdot \sigma^4} - 3$$

$$\text{Pearson's Second Coefficient of Kurtosis:} \quad \text{Kurtosis} = \frac{\frac{1}{n}\sum_{i=1}^{n}(x_i - \bar{x})^4}{\left(\frac{1}{n}\sum_{i=1}^{n}(x_i - \bar{x})^2\right)^2} - 3$$

$$\text{Excess Kurtosis:} \quad \text{Excess Kurtosis} = \text{Kurtosis} - 3$$

These measures provide insights into the shape and distribution of a dataset, capturing skewness and kurtosis.

2.5   Graphical Representation

Graphical representation visually conveys data trends and patterns. Key concepts include:

- **Histogram:** Visualize distribution of data. Bin data into intervals and create bars representing the frequency.

- **Box Plot:** Illustrates the five-number summary - minimum, lower quartile, median, upper quartile, and maximum.

- **Scatter Plot:** Displays relationships between two variables. Each point represents a data pair.

- **Line Chart:** Connects data points with straight lines, ideal for showing trends over time or continuous data.

- **Pie Chart:** Represents parts of a whole. Each sector corresponds to a percentage of the total.

- **Probability Density Function (PDF):** Graphical representation of a continuous probability distribution.

In LaTeX, you can include graphics using the `graphicx` package:

`\includegraphics[width=0.8\textwidth]{your_image_file.png}`

These graphical representations enhance data interpretation and analysis.

2.6   Summary and Exercises

Summary

Recap of Key Formulas and Concepts:

$$\text{Measures of Central Tendency:} \quad \bar{x}, \text{ Median, Mode}$$

$$\text{Measures of Dispersion:} \quad \sigma, \text{ IQR, Range}$$

$$\text{Percentiles and Quartiles:} \quad P_k, Q_1, Q_2, Q_3$$

$$\text{Skewness and Kurtosis:} \quad \text{Skewness, Kurtosis}$$

$$\text{Graphical Representation:} \quad \text{Histogram, Box Plot, Scatter Plot}$$

Exercises

Apply what you've learned:

1. Calculate the mean and standard deviation for a given dataset. 2. Construct a box plot for a set of values. 3. Compute the interquartile range (IQR) for a dataset. 4. Determine the skewness and kurtosis of a distribution. 5. Create a histogram for a given frequency distribution.

Challenge

Explore advanced topics:

Multivariate Analysis: Covariance Matrix, Principal Component Analysis (PCA)

Bayesian Statistics: Bayesian Inference, Posterior Probability

Machine Learning: Linear Regression, Decision Trees, Clustering Algorithms

Explore these topics to deepen your understanding.

Chapter 3

Probability Distributions

3.1 Discrete Distributions

This section introduces key discrete probability distributions with memorable mathematical expressions:

Probability Mass Function (PMF)

$$P(X = x) = f(x)$$

Bernoulli Distribution:

$$f(x; p) = p^x (1-p)^{1-x} \quad \text{for} \quad x \\ \in \{0, 1\}$$

Binomial Distribution:

$$f(x; n, p) = \binom{n}{x} p^x (1-p)^{n-x} \quad \text{for} \quad x \\ \in \{0, 1, 2, \ldots, n\}$$

Poisson Distribution:

$$f(x; \lambda) = \frac{e^{-\lambda} \lambda^x}{x!} \quad \text{for} \quad x \\ \in \{0, 1, 2, \ldots\}$$

11

Cumulative Distribution Function (CDF)

$$F(x) = P(X \leq x)$$

Bernoulli CDF:

$$F(x; p) = \begin{cases} 0 & \text{for} \quad x < 0 \\ 1 - p & \text{for} \quad 0 \leq x < 1 \\ 1 & \text{for} \quad x \geq 1 \end{cases}$$

Binomial CDF:

$$F(x; n, p) = \sum_{k=0}^{x} \binom{n}{k} p^k (1-p)^{n-k}$$

Poisson CDF:

$$F(x; \lambda) = e^{-\lambda} \sum_{k=0}^{x} \frac{\lambda^k}{k!}$$

These distributions model random events and their probabilities in a discrete context.

3.2 Continuous Distributions

This section introduces key concepts related to continuous probability distributions in a fast and memorable way:

$$\text{Probability Density Function (PDF):} \quad f(x) \geq 0, \quad \int_{-\infty}^{\infty} f(x)\, dx = 1$$

$$\text{Cumulative Distribution Function (CDF):} \quad F(x) = \int_{-\infty}^{x} f(t)\, dt$$

$$\text{Expected Value (Mean):} \quad \mu = \int_{-\infty}^{\infty} x \cdot f(x)\, dx$$

$$\text{Variance:} \quad \sigma^2 = \int_{-\infty}^{\infty} (x - \mu)^2 \cdot f(x)\, dx$$

$$\text{Standard Deviation:} \quad \sigma = \sqrt{\sigma^2}$$

$$\text{Normal Distribution:} \quad f(x|\mu, \sigma^2) = \frac{1}{\sqrt{2\pi\sigma^2}} e^{-\frac{(x-\mu)^2}{2\sigma^2}}$$

$$\text{Uniform Distribution:} \quad f(x|a, b) = \frac{1}{b-a}, \quad a \le x \le b$$

These formulas define the fundamentals of continuous probability distributions, providing insights into random variables and their properties.

3.3 Probability Density Functions

This section introduces key concepts related to probability density functions (PDFs) in a fast and memorable way:

$$\text{Probability Density Function (PDF):} \quad f(x)$$
$$\ge 0, \quad \int_{-\infty}^{\infty} f(x)\, dx$$
$$= 1$$

$$\text{Cumulative Distribution Function (CDF):} \quad F(x) = \int_{-\infty}^{x} f(t)\, dt$$

$$\text{Expected Value (Mean):} \quad \mu = \int_{-\infty}^{\infty} x \cdot f(x)\, dx$$

$$\text{Variance:} \quad \sigma^2 = \int_{-\infty}^{\infty} (x - \mu)^2 \cdot f(x)\, dx$$

$$\text{Standard Deviation:} \quad \sigma = \sqrt{\sigma^2}$$

$$\text{Normal Distribution:} \quad f(x|\mu, \sigma^2) = \frac{1}{\sqrt{2\pi\sigma^2}} e^{-\frac{(x-\mu)^2}{2\sigma^2}}$$

$$\text{Uniform Distribution:} \quad f(x|a, b) = \frac{1}{b-a}, \quad a \le x \le b$$

These formulas encapsulate the essence of probability density functions, offering a foundation for understanding continuous probability distributions.

3.4   Cumulative Distribution Functions

This section introduces essential concepts related to cumulative distribution functions (CDFs) in a quick and memorable way:

$$\text{CDF:}\quad F(x) = \int_{-\infty}^{x} f(t)\, dt$$

$$\text{PDF:}\quad f(x) \geq 0,\quad \int_{-\infty}^{\infty} f(x)\, dx = 1$$

$$\text{Mean (Expected Value):}\quad \mu = \int_{-\infty}^{\infty} x \cdot f(x)\, dx$$

$$\text{Variance:}\quad \sigma^2 = \int_{-\infty}^{\infty} (x - \mu)^2 \cdot f(x)\, dx$$

$$\text{Standard Deviation:}\quad \sigma = \sqrt{\sigma^2}$$

$$\text{Normal Distribution:}\quad f(x|\mu, \sigma^2) = \frac{1}{\sqrt{2\pi\sigma^2}} e^{-\frac{(x-\mu)^2}{2\sigma^2}}$$

$$\text{Uniform Distribution:}\quad f(x|a, b) = \frac{1}{b-a},\quad a \leq x \leq b$$

These formulas encapsulate the essence of cumulative distribution functions, providing a foundation for understanding probability distributions.

3.5   Expected Values

This section introduces key concepts related to expected values in a quick and memorable way:

$$\text{Expected Value (Mean):}\quad \mu = \int_{-\infty}^{\infty} x \cdot f(x)\, dx$$

$$\text{Variance:}\quad \sigma^2 = \int_{-\infty}^{\infty} (x - \mu)^2 \cdot f(x)\, dx$$

$$\text{Standard Deviation:} \quad \sigma = \sqrt{\sigma^2}$$

$$\text{Linearity of Expectation:} \quad E[aX + bY] = aE[X] + bE[Y]$$

$$\text{Expected Value of a Constant:} \quad E[c] = c$$

$$\text{Expected Value of a Random Variable:} \quad E[g(X)] = \int_{-\infty}^{\infty} g(x) \cdot f(x)\, dx$$

These formulas provide a succinct overview of expected values and their properties, forming a basis for understanding probability and statistics.

3.6 Moment Generating Functions

This section introduces moment generating functions (MGFs) in a quick and memorable way:

$$\text{Moment Generating Function (MGF):} \quad M_X(t) = \mathbb{E}\left[e^{tX}\right]$$

$$\text{Derivative at 0:} \quad M'_X(0) = \mathbb{E}[X]$$

$$\text{Second Derivative at 0:} \quad M''_X(0) = \mathbb{E}[X^2]$$

$$\text{Nth Derivative at 0:} \quad M_X^{(n)}(0) = \mathbb{E}[X^n]$$

$$\text{Sum of Independent Random Variables:} \quad M_{X+Y}(t) = M_X(t) \cdot M_Y(t)$$

$$\text{Linear Transformation:} \quad M_{aX+b}(t) = e^{bt} \cdot M_X(at)$$

MGFs offer a powerful tool for computing expected values and moments, especially for sums of independent random variables and linear transformations.

Chapter 4

Sampling Theory

4.1 Random Sampling

This section introduces key concepts related to random sampling in a quick and memorable way:

$$\text{Population Mean:} \quad \mu = \frac{1}{N} \sum_{i=1}^{N} x_i$$

$$\text{Sample Mean:} \quad \bar{x} = \frac{1}{n} \sum_{i=1}^{n} x_i$$

$$\text{Population Variance:} \quad \sigma^2 = \frac{1}{N} \sum_{i=1}^{N} (x_i - \mu)^2$$

$$\text{Sample Variance:} \quad s^2 = \frac{1}{n-1} \sum_{i=1}^{n} (x_i - \bar{x})^2$$

$$\text{Sampling Distribution of the Mean:} \quad \bar{x} \sim N\left(\mu, \frac{\sigma^2}{n}\right)$$

Central Limit Theorem: $\bar{x}$ approaches normal distribution for sufficiently large n

Random sampling allows us to make inferences about a population based on characteristics observed in a sample.

4.2   Sampling Distributions

Key concepts related to sampling distributions presented in a quick and memorable way:

$$\text{Population Mean:} \quad \mu = \frac{1}{N} \sum_{i=1}^{N} x_i$$

$$\text{Sample Mean:} \quad \bar{x} = \frac{1}{n} \sum_{i=1}^{n} x_i$$

$$\text{Population Variance:} \quad \sigma^2 = \frac{1}{N} \sum_{i=1}^{N} (x_i - \mu)^2$$

$$\text{Sample Variance:} \quad s^2 = \frac{1}{n-1} \sum_{i=1}^{n} (x_i - \bar{x})^2$$

$$\text{Sampling Distribution of the Mean:} \quad \bar{x} \sim N\left(\mu, \frac{\sigma^2}{n}\right)$$

Central Limit Theorem: $\bar{x}$ approaches normal distribution for sufficiently large n

Sampling distributions provide insights into the behavior of sample statistics and enable meaningful population inferences.

4.3   Estimation

This section introduces key concepts related to estimation in a quick and memorable way:

$$\text{Point Estimation:} \quad \hat{\theta} = g(x_1, x_2, ..., x_n)$$

$$\text{Interval Estimation:} \quad \hat{\theta}_{\text{lower}} \leq \theta \\ \leq \hat{\theta}_{\text{upper}}$$

$$\text{Bias of an Estimator:} \quad \text{Bias}(\hat{\theta}) = \mathbb{E}(\hat{\theta}) - \theta$$

$$\text{Mean Squared Error (MSE):} \quad \text{MSE}(\hat{\theta}) = \text{Var}(\hat{\theta}) + \left(\text{Bias}(\hat{\theta})\right)^2$$

$$\text{Confidence Interval Width:} \quad \text{Width} = 2 \times \text{Margin of Error}$$

$$\text{Margin of Error:} \quad \text{Margin of Error} = \frac{Z \times \sigma}{\sqrt{n}}$$

Estimation techniques play a crucial role in statistical analysis, providing insights into population parameters.

4.4 Confidence Intervals

This section introduces key concepts related to confidence intervals in a quick and memorable way:

$$\text{Confidence Interval (CI):} \quad \hat{\theta} \pm Z \times \frac{\sigma}{\sqrt{n}}$$

$$\text{Confidence Level:} \quad \text{CI} = \hat{\theta} \pm \text{Margin of Error}$$

$$\text{Margin of Error:} \quad \text{Margin of Error} = Z \times \frac{\sigma}{\sqrt{n}}$$

$$\text{Standard Error:} \quad \text{SE} = \frac{\sigma}{\sqrt{n}}$$

$$\text{Z-Score:} \quad Z \text{ is the critical value based on the desired confidence level}$$

$$\text{Interpretation:} \quad \text{CI} \left(\hat{\theta} \pm \text{Margin of Error}\right) \text{ with confidence level } \%$$

Confidence intervals provide a range of values within which we can reasonably expect the true parameter to lie.

4.5 Hypothesis Testing

Key concepts related to hypothesis testing presented in a quick and memorable way:

$$\text{Null Hypothesis:} \quad H_0 : \theta$$
$$= \theta_0$$

$$\text{Alternative Hypothesis:} \quad H_1 \text{ (or } H_a\text{):} \quad \theta$$
$$\neq \theta_0 \text{ (Two-Tailed)} \quad \text{or} \quad \theta$$
$$> \theta_0 \text{ (One-Tailed)}$$

$$\text{Test Statistic:} \quad z = \frac{\hat{\theta} - \theta_0}{\text{SE}(\hat{\theta})}$$

$$\text{P-Value:} \quad \text{P-Value} = P(Z$$
$$> |z|) \text{ (Two-Tailed)} \quad \text{or} \quad \text{P-Value}$$
$$= P(Z$$
$$> z) \text{ (One-Tailed)}$$

$$\text{Decision Rule:} \quad \text{Reject } H_0 \text{ if P-Value } \leq \alpha \text{ (Significance Level)}$$

$$\text{Type I Error:} \quad \alpha = P(\text{Reject } H_0 \,|\, H_0 \text{ is true})$$

$$\text{Type II Error:} \quad \beta = P(\text{Fail to reject } H_0 \,|\, H_1 \text{ is true})$$

Hypothesis testing facilitates decision-making about population parameters using sample data.

4.6 Power and Sample Size Calculations

This section introduces key concepts related to power and sample size calculations in a quick and memorable way:

$$\text{Power of a Test:} \quad \text{Power} = 1 - \beta$$

$$\text{Critical Value:} \quad z_{\alpha/2} \text{ (for a two-tailed test)}$$

$$\text{Effect Size:} \quad \text{Effect Size} = \frac{\text{Difference in Means}}{\text{Standard Deviation}}$$

$$\text{Sample Size Formula:} \quad n = \left(\frac{Z_{\alpha/2} + Z_\beta}{\text{Effect Size}} \right)^2$$

$$\text{Critical Region:} \quad \text{Reject } H_0 \text{ if } z > z_{\alpha/2} + \text{Effect Size} \times Z_\beta$$

Power and sample size calculations are crucial for designing experiments with sufficient sensitivity to detect meaningful effects.

Chapter 5

Regression Analysis

5.1　Simple Linear Regression

This section introduces key concepts related to simple linear regression in a quick and memorable way:

Simple Linear Regression Model:　$Y = \beta_0 + \beta_1 X + \varepsilon$

Least Squares Estimation:　$\hat{\beta}_1 = \dfrac{\sum_{i=1}^{n}(X_i - \bar{X})(Y_i - \bar{Y})}{\sum_{i=1}^{n}(X_i - \bar{X})^2}$

Intercept Estimate:　$\hat{\beta}_0 = \bar{Y} - \hat{\beta}_1 \bar{X}$

Regression Equation:　$\hat{Y} = \hat{\beta}_0 + \hat{\beta}_1 X$

Residuals:　$e_i = Y_i - \hat{Y}_i$

Coefficient of Determination:　$R^2 = \dfrac{\sum_{i=1}^{n}(\hat{Y}_i - \bar{Y})^2}{\sum_{i=1}^{n}(Y_i - \bar{Y})^2}$

Simple linear regression helps model the relationship between two variables and make predictions based on observed data.

5.2   Multiple Linear Regression

This section introduces key concepts related to multiple linear regression in a quick and memorable way:

$$\text{Model:} \quad Y = \beta_0 + \beta_1 X_1 + \beta_2 X_2 + \ldots + \beta_p X_p + \varepsilon$$

$$\text{Vector Notation:} \quad \mathbf{Y} = \beta_0 + \beta_1 \mathbf{X}_1 + \beta_2 \mathbf{X}_2 + \ldots + \beta_p \mathbf{X}_p + \text{"}$$

$$\text{Matrix Notation:} \quad \mathbf{Y} = \mathbf{XB} + \text{"}$$

$$\text{Least Squares Estimation:} \quad \hat{\mathbf{B}} = (\mathbf{X}^T\mathbf{X})^{-1}\mathbf{X}^T\mathbf{Y}$$

$$\text{Coefficient of Determination:} \quad R^2 = \frac{\text{SSR}}{\text{SSTO}}$$
$$= 1 - \frac{\text{SSE}}{\text{SSTO}}$$

$$\text{Adjusted } R^2: \quad \text{Adjusted } R^2 = 1 - \left(\frac{\text{SSE}/(n-p-1)}{\text{SSTO}/(n-1)} \right)$$

Multiple linear regression allows us to model the relationship between a dependent variable and multiple independent variables.

5.3   Logistic Regression

This section introduces key concepts related to logistic regression in a quick and memorable way:

$$\text{Model:} \quad \ln\left(\frac{p}{1-p} \right) = \beta_0 + \beta_1 X_1 + \beta_2 X_2 + \ldots + \beta_p X_p$$

$$\text{Odds Ratio:} \quad \text{Odds}(Y = 1) = e^{\beta_0 + \beta_1 X_1 + \beta_2 X_2 + \ldots + \beta_p X_p}$$

$$\text{Probability:} \quad P(Y = 1) = \frac{e^{\beta_0 + \beta_1 X_1 + \beta_2 X_2 + \ldots + \beta_p X_p}}{1 + e^{\beta_0 + \beta_1 X_1 + \beta_2 X_2 + \ldots + \beta_p X_p}}$$

$$\text{Logit Function:} \quad \text{logit}(p) = \ln\left(\frac{p}{1-p} \right)$$

Maximum Likelihood Estimation: Maximize $\mathcal{L}(\beta)$

$$= \prod_{i=1}^{n} P(Y_i$$
$$= y_i)^{y_i} \times [1 - P(Y_i$$
$$= y_i)]^{1-y_i}$$

Receiver Operating Characteristic (ROC) Curve: Graph of Sensitivity vs. (1-Specificity)

Logistic regression is valuable for modeling binary outcomes and estimating the probability of an event occurring.

5.4   Nonlinear Regression

This section introduces key concepts related to nonlinear regression in a quick and memorable way:

$$\text{Model:} \quad Y = f(X, \beta) + \varepsilon$$

$$\text{Parameter Estimation:} \quad \hat{\beta} = \arg\min_{\beta} \sum_{i=1}^{n} (Y_i - f(X_i, \beta))^2$$

$$\text{Common Models:} \quad Y = \beta_0 e^{\beta_1 X} + \varepsilon \quad \text{or} \quad Y$$
$$= \frac{\beta_0}{1 + e^{-\beta_1 X}} + \varepsilon$$

$$\text{Gauss-Newton Algorithm:} \quad \beta^{(k+1)} = \beta^{(k)} + \left(J^T J\right)^{-1} J^T \mathbf{r}(\beta^{(k)})$$

$$\text{Residuals:} \quad \mathbf{r}(\beta) = \begin{bmatrix} Y_1 - f(X_1, \beta) \\ Y_2 - f(X_2, \beta) \\ \vdots \\ Y_n - f(X_n, \beta) \end{bmatrix}$$

$$\text{Model Assessment:} \quad \text{R-squared, AIC, BIC, etc.}$$

Nonlinear regression is employed when the relationship between variables is not linear, and it involves estimating parameters to fit a given model to the data.

5.5   Model Checking and Diagnostics

This section introduces key concepts related to model checking and diagnostics in a quick and memorable way:

$$\text{Residuals:} \quad \text{Residuals} = \text{Observed Values} - \text{Fitted Values}$$

$$\text{Residual Analysis:} \quad \text{Check for patterns, outliers, and non-constant variance}$$

$$\text{Normal Probability Plot:} \quad \text{Plot of residuals against theoretical quantiles}$$

$$\text{Cook's Distance:} \quad D_i = \frac{\sum_{j=1}^{n} (\hat{Y}_j - \hat{Y}_{j(i)})^2}{p \times \text{MSE}}$$

$$\text{Influence Plot:} \quad \text{Visualize the impact of each observation on the regression coefficients}$$

$$\text{Multicollinearity:} \quad \text{Variance Inflation Factor (VIF)}$$

$$\text{Heteroscedasticity:} \quad \text{Breusch-Pagan Test}$$

Model checking and diagnostics help ensure the reliability of regression models and identify potential issues for refinement.

5.6   Variable Selection

This section introduces key concepts related to variable selection in a quick and memorable way:

$$\text{Forward Selection:} \quad \text{Start with an empty model, add variables one by one}$$

$$\text{Backward Elimination:} \quad \text{Start with all variables, remove one at a time}$$

$$\text{Stepwise Selection:} \quad \text{Combination of forward and backward steps}$$

$$\text{AIC (Akaike Information Criterion):} \quad \text{AIC} = 2k - 2\ln\left(\hat{L}\right)$$

$$\text{BIC (Bayesian Information Criterion):} \quad \text{BIC} = k\ln(n) - 2\ln\left(\hat{L}\right)$$

$$\text{Cross-Validation:} \quad \text{Divide data into training and testing sets, assess model performance}$$

$$\text{Regularization:} \quad \text{LASSO (Least Absolute Shrinkage and Selection Operator), Ridge Regression}$$

Variable selection methods help identify the most influential variables for an effective and parsimonious model.

Chapter 6

Statistical Inference

6.1 Parametric Inference

This section introduces key concepts related to parametric inference in a quick and memorable way:

$$\text{Population Parameter:} \quad \theta$$

$$\text{Point Estimation:} \quad \hat{\theta} = g(X_1, X_2, ..., X_n)$$

$$\text{Confidence Interval:} \quad \hat{\theta} \pm z_{\alpha/2} \times \text{SE}(\hat{\theta})$$

$$\text{Hypothesis Testing:} \quad H_0 : \theta = \theta_0 \quad \text{vs.} \quad H_1 : \theta \neq \theta_0$$

$$\text{Test Statistic:} \quad z = \frac{\hat{\theta} - \theta_0}{\text{SE}(\hat{\theta})}$$

$$\text{P-Value:} \quad \text{P-Value} = P(|Z| > |z|) \quad \text{(Two-Tailed)}$$

$$\text{Type I Error:} \quad \alpha = P(\text{Reject } H_0 \mid H_0 \text{ is true})$$

Parametric inference involves making inferences about population parameters based on sample data using estimation and hypothesis testing techniques.

6.2 Nonparametric Inference

This section introduces key concepts related to nonparametric inference in a quick and memorable way:

$$\text{Population Distribution:} \quad F(x)$$

$$\text{Ranking of Observations:} \quad R_1, R_2, ..., R_n$$

$$\text{Wilcoxon Signed-Rank Test:} \quad T^+ = \min\left(\sum_{i=1}^{n} \text{sign}(X_i) \times R_i, \sum_{i=1}^{n} \text{sign}(X_i - 1) \times R_i\right)$$

$$\text{Mann-Whitney U Test:} \quad U = R_1 - \frac{n_1(n_1 + 1)}{2}$$

$$\text{Kolmogorov-Smirnov Test:} \quad D = \max_{i}\left(\frac{i}{n} - F(X_{(i)}), F(X_{(i)}) - \frac{i-1}{n}\right)$$

$$\text{Runs Test:} \quad R = \text{Number of Runs}$$

$$\text{Bootstrap Confidence Interval:} \quad \text{Calculate percentiles from bootstrap samples}$$

Nonparametric inference methods provide alternatives for analyzing data without assuming specific population distributions.

6.3 Bootstrap Methods

This section introduces key concepts related to bootstrap methods in a quick and memorable way:

$$\text{Resampling Procedure:} \quad \text{Draw } B \text{ bootstrap samples from the observed data}$$

$$\text{Bootstrap Sample:} \quad X_1^*, X_2^*, ..., X_n^*$$

$$\text{Bootstrap Estimate:} \quad \hat{\theta}^* = g(X_1^*, X_2^*, ..., X_n^*)$$

$$\text{Bootstrap Bias:} \quad \text{Bias}(\hat{\theta}^*) = \hat{\theta}^* - \hat{\theta}$$

$$\text{Bootstrap Standard Error:} \quad \text{SE}(\hat{\theta}^*) = \sqrt{\frac{1}{B-1} \sum_{b=1}^{B} (\hat{\theta}_b^* - \bar{\hat{\theta}}^*)^2}$$

Bootstrap Confidence Interval: Calculate percentiles from bootstrap estimates

Bootstrap Hypothesis Test: Calculate the proportion of bootstrap estimates more extreme than the observed value

Bootstrap methods provide a powerful tool for estimating the sampling distribution of a statistic and making inferences from a sample.

6.4   Bayesian Inference

This section introduces key concepts related to Bayesian inference in a quick and memorable way:

$$\text{Bayesian Framework:} \quad \text{Posterior} \propto \text{Likelihood} \times \text{Prior}$$

$$\text{Posterior Distribution:} \quad \pi(\theta|\mathbf{X}) \propto f(\mathbf{X}|\theta) \times \pi(\theta)$$

$$\text{Prior Distribution:} \quad \pi(\theta) \quad (\text{Prior beliefs about } \theta)$$

$$\text{Likelihood Function:} \quad f(\mathbf{X}|\theta) \quad (\text{Probability of the data given } \theta)$$

$$\text{Posterior Mean:} \quad \text{E}(\theta|\mathbf{X}) = \int \theta \times \pi(\theta|\mathbf{X}) \, d\theta$$

$$\text{Posterior Variance:} \quad \text{Var}(\theta|\mathbf{X}) = \int (\theta - \text{E}(\theta|\mathbf{X}))^2 \times \pi(\theta|\mathbf{X}) \, d\theta$$

$$\text{Bayesian Credible Interval:} \quad [\text{5th Percentile}, \text{95th Percentile}]$$

Bayesian inference provides a flexible framework for incorporating prior knowledge into statistical analysis and updating beliefs based on observed data.

6.5   Statistical Decision Theory

This section introduces key concepts related to statistical decision theory in a quick and memorable way:

$$\text{Decision Space:} \quad \mathcal{D} = \{\text{Accept } H_0, \ \text{Reject } H_0\}$$

$$\text{Loss Function:} \quad L(\theta, a) \quad (\text{Loss incurred for decision } a \text{ when the true state is } \theta)$$

$$\text{Risk Function:} \quad R(a) = \int_{\Theta} L(\theta, a) \times \pi(\theta) \, d\theta \quad (\text{Expected loss})$$

$$\text{Decision Rule:} \quad \delta(\mathbf{X}) = \begin{cases} \text{Accept } H_0 & \text{if } R(\text{Accept } H_0) < R(\text{Reject } H_0) \\ \text{Reject } H_0 & \text{if } R(\text{Reject } H_0) < R(\text{Accept } H_0) \end{cases}$$

Example: Consider a medical test for a rare disease.
  Accept H_0 (no disease) may have a small cost, but Reject H_0
   (incorrectly diagnosing disease
  when it's not present) may have a much higher cost.
  The loss function helps quantify these costs.

$$\text{Bayes' Decision Rule:} \quad \delta_{\text{Bayes}}(\mathbf{X}) = \begin{cases} \text{Accept } H_0 & \text{if } \pi(H_0|\mathbf{X}) > \pi(H_1|\mathbf{X}) \\ \text{Reject } H_0 & \text{otherwise} \end{cases}$$

Example: In drug testing, $\pi(H_0|\mathbf{X})$ is the probability that the drug is safe, and $\pi(H_1|\mathbf{X})$ is the probability that the drug is harmful.
The decision is based on these probabilities.

$$\text{Minimax Decision Rule:} \quad \delta_{\text{Minimax}}(\mathbf{X}) = \arg \min_{a \in \mathcal{D}} \max_{\theta \in \Theta} R(a)$$

Example: In a legal case, the minimax decision rule aims to minimize the maximum possible loss, considering the worst-case scenario for each decision.

  Statistical decision theory provides a systematic framework for making optimal decisions in the face of uncertainty, balancing the costs of errors and incorporating prior information.

6.6   Robust Statistics

This section introduces key concepts related to robust statistics in a quick and memorable way:

Objective: Minimize the impact of outliers on statistical estimates

Median: Robust measure of central tendency $\tilde{X} = \text{median}(X)$

MAD (Median Absolute Deviation):   Measure of dispersion   MAD $= \text{median}(|X - \tilde{X}|)$

Trimmed Mean: Remove a certain percentage of extreme values $\bar{X}_\alpha$

$$= \frac{1}{n(1 - \alpha)} \sum_{i=\alpha n+1}^{(1-\alpha)n} X_{(i)}$$

Winsorizing: Replace extreme values with the nearest non-extreme value

Huber's M-Estimator: $Q_\rho(c) = \begin{cases} \frac{1}{2}c^2 & \text{if } |c| \le \rho \\ \rho(|c| - \frac{1}{2}\rho) & \text{otherwise} \end{cases}$

Example: Consider the dataset $[2, 5, 8, 10, 1000]$
Median: $\tilde{X} = 8$ MAD: MAD $= 1$
Trimmed Mean (10%): $\bar{X}_{0.1}$ $= 6.25$

Robust statistics focuses on providing reliable estimates even in the presence of outliers or skewed distributions, reducing their influence on the analysis.

Chapter 7

Time Series Analysis

7.1   Time Series Components

This section introduces key concepts related to time series components in a quick and memorable way:

$$\text{Trend Component:}\quad Y_t = T_t + \varepsilon_t$$

$$\text{Seasonal Component:}\quad Y_t = T_t + S_t + \varepsilon_t$$

$$\text{Cyclic Component:}\quad Y_t = T_t + S_t + C_t + \varepsilon_t$$

$$\text{Irregular or Residual Component:}\quad Y_t = T_t + S_t + C_t + \varepsilon_t$$

Example: Consider monthly sales data for a retail store

Assume a linear trend, a seasonal pattern, and random fluctuations

$$\text{Model:}\quad Y_t = 100 + 5t + 20\sin\left(\frac{2\pi t}{12}\right) + \varepsilon_t$$

Time series components help to decompose observed data into underlying patterns, facilitating better understanding and forecasting.

7.2   Autoregressive Integrated Moving Average (ARIMA)

This section introduces key concepts related to ARIMA models in a quick and memorable way:

ARIMA(p, d, q) Model: $(1 - \phi_1 B - \phi_2 B^2 - \ldots - \phi_p B^p)(1 - B)^d Y_t$
$= (1 + \theta_1 B + \theta_2 B^2 + \ldots + \theta_q B^q)\varepsilon_t$

Autoregressive (AR) Component: $\phi_1, \phi_2, \ldots, \phi_p$

Integrated (I) Component: d (Order of differencing)

Moving Average (MA) Component: $\theta_1, \theta_2, \ldots, \theta_q$

Example: Consider monthly temperature data
Model: $(1 - 0.7B)(1 - B)(Y_t - 25) = (1 + 0.4B)\varepsilon_t$

ARIMA models are powerful tools for time series forecasting, capturing trends, seasonality, and random fluctuations.

7.3   Seasonal Decomposition of Time Series (STL)

This section introduces key concepts related to STL decomposition in a quick and memorable way:

STL Decomposition: $Y_t = T_t + S_t + C_t + \varepsilon_t$

Trend Component (T): T_t (Smoothed trend)

Seasonal Component (S): S_t (Periodic pattern)

Residual Component (C): C_t (Irregular fluctuations)

Example: Consider monthly sales data for a retail store
Apply STL decomposition to reveal underlying trends and seasonality

STL decomposition is a powerful technique for extracting underlying patterns from time series data, aiding in trend analysis and forecasting.

7.4   Forecasting Methods

This section introduces key forecasting methods in a quick and memorable way:

$$\text{Moving Average (MA) Forecast:} \quad \hat{Y}_{t+1} = \frac{1}{m} \sum_{i=0}^{m-1} Y_{t-i}$$

$$\text{Exponential Smoothing (ES) Forecast:} \quad \hat{Y}_{t+1} = \alpha Y_t + (1 - \alpha)\hat{Y}_t$$

$$\text{ARIMA Forecast:} \quad \hat{Y}_{t+1} = \phi_1 Y_t + \phi_2 Y_{t-1} + \ldots + \phi_p Y_{t-p} + \varepsilon_{t+1}$$

$$\text{STL Forecast:} \quad \hat{Y}_{t+1} = \hat{T}_{t+1} + \hat{S}_{t+1} + \hat{C}_{t+1}$$

Example: Forecast monthly sales for the next period using the given data and method

Forecasting methods provide valuable insights into future trends and patterns, aiding in decision-making and planning.

7.5   Model Evaluation

This section introduces key concepts related to model evaluation in a quick and memorable way:

$$\text{Mean Squared Error (MSE):} \quad \text{MSE} = \frac{1}{n} \sum_{i=1}^{n} (Y_i - \hat{Y}_i)^2$$

$$\text{Mean Absolute Error (MAE):} \quad \text{MAE} = \frac{1}{n} \sum_{i=1}^{n} |Y_i - \hat{Y}_i|$$

$$\text{Root Mean Squared Error (RMSE):} \quad \text{RMSE} = \sqrt{\text{MSE}}$$

$$\text{Percentage Error (PE):} \quad \text{PE} = \frac{1}{n} \sum_{i=1}^{n} \left| \frac{Y_i - \hat{Y}_i}{Y_i} \right| \times 100$$

Example: Evaluate the performance of a forecasting model using given data and calculate MSE, MAE, RMSE, and PE

Model evaluation metrics help assess the accuracy and reliability of predictive models, guiding improvements and optimizations.

7.6   Financial Time Series Analysis

This section introduces key concepts related to financial time series analysis in a quick and memorable way:

$$\text{Logarithmic Returns:} \quad r_t = \ln\left(\frac{P_t}{P_{t-1}}\right)$$

$$\text{Volatility:} \quad \sigma_t = \sqrt{\frac{1}{n-1}\sum_{i=1}^{n}(r_i - \bar{r})^2}$$

$$\text{Moving Average Convergence Divergence (MACD):} \quad \text{MACD} = EMA_{12} - EMA_{26}$$

$$\text{Relative Strength Index (RSI):} \quad \text{RSI} = 100 - \frac{100}{1 + \text{RS}}$$

Example: Analyze daily stock prices using logarithmic returns, calculate volatility, MACD, and RSI.

Financial time series analysis helps investors and analysts make informed decisions by identifying trends, measuring risk, and assessing market momentum.

Chapter 8

Experimental Design

8.1 Randomized Designs

This section introduces key concepts related to randomized designs in a quick and memorable way:

Randomized Complete Block Design (RCBD): $Y_{ij} = \mu + \tau_i + \beta_j + \varepsilon_{ij}$

Latin Square Design: $Y_{ijk} = \mu + \alpha_i + \beta_j + \gamma_k + \varepsilon_{ijk}$

Factorial Design: $Y_{ijk} = \mu + \tau_i + \beta_j + \gamma_k + (\tau\beta)_{ij} + (\tau\gamma)_{ik} + (\beta\gamma)_{jk} + \varepsilon_{ijk}$

Example: Conduct an experiment to test the effects of fertilizer types on crop yield using a randomized complete block design

Randomized designs in experimental studies help control for potential confounding variables and provide more robust and reliable results.

8.2 Block Designs

This section introduces key concepts related to block designs in a quick and memorable way:

Block Design Model: $Y_{ijk} = \mu + \alpha_i + \beta_j + (\alpha\beta)_{ij} + \gamma_k + (\alpha\gamma)_{ik}$
$$+ (\beta\gamma)_{jk} + (\alpha\beta\gamma)_{ijk} + \varepsilon_{ijk}$$

Completely Randomized Design (CRD): $Y_{ij} = \mu + \alpha_i + \varepsilon_{ij}$

Randomized Complete Block Design (RCBD): $Y_{ijk} = \mu + \alpha_i + \beta_j + (\alpha\beta)_{ij} + \varepsilon_{ijk}$

Example: Investigate the effect of different soil types on plant growth using a randomized complete block design

Block designs in experimental studies help control for variability and confounding factors, improving the precision and reliability of results.

8.3   Factorial Designs

This section introduces key concepts related to factorial designs in a quick and memorable way:

Factorial Design Model: $Y_{ijk} = \mu + \tau_i + \beta_j + (\tau\beta)_{ij} + \gamma_k + (\tau\gamma)_{ik}$
$$+ (\beta\gamma)_{jk} + (\tau\beta\gamma)_{ijk} + \varepsilon_{ijk}$$

$$\text{Main Effects:}\quad \text{ME}_{\tau_i} = \frac{1}{2^b} \sum_{j=1}^{2^b} Y_{ij\ldots k}$$

$$\text{Interaction Effects:}\quad \text{IE}_{\tau_i \beta_j} = \frac{1}{2^b} \sum_{k=1}^{2^c} Y_{ijk\ldots}$$

Example: Study the impact of varying levels of temperature and humidity on crop yield using a factorial design

Factorial designs allow researchers to study the effects of multiple factors and their interactions simultaneously, providing a comprehensive understanding of experimental outcomes.

8.4   Latin Square Designs

This section introduces key concepts related to Latin square designs in a quick and memorable way:

Latin Square Model: $Y_{ijk} = \mu + \alpha_i + \beta_j + \gamma_k + \varepsilon_{ijk}$

Balanced Latin Square: Each treatment occurs exactly once in each row and each column

$$\text{Orthogonality Condition:} \quad \sum_{i=1}^{n} \alpha_i$$

$$= \sum_{j=1}^{n} \beta_j$$

$$= \sum_{k=1}^{n} \gamma_k$$

$$= 0$$

Example: Investigate the impact of different fertilizers on crop yield using a Latin square design

Latin square designs provide a structured approach to control for variability and systematic errors, enhancing the validity of experimental results.

8.5   Crossover Designs

This section introduces key concepts related to crossover designs in a quick and memorable way:

$$\text{Crossover Model:} \quad Y_{ij} = \mu + \alpha_i + \beta_j + \varepsilon_{ij}$$

Period Effect: α_i (Effect of the ith period)

Treatment Effect: β_j (Effect of the jth treatment)

Washout Period: A break or washout period between treatment sequences

Example: Assess the efficacy of two medications with a crossover design involving multiple treatment periods

Crossover designs allow for within-subject comparisons and help control for individual variability, providing more robust results in clinical trials and medical studies.

8.6   Analysis of Variance (ANOVA)

This section introduces key concepts related to analysis of variance in a quick and memorable way:

$$\text{One-Way ANOVA Model:} \quad Y_{ij} = \mu + \tau_i + \varepsilon_{ij}$$

$$\text{Two-Way ANOVA Model:} \quad Y_{ijk} = \mu + \alpha_i + \beta_j + (\alpha\beta)_{ij} + \varepsilon_{ijk}$$

$$\text{Total Sum of Squares (SST):} \quad \text{SST} = \sum_{i=1}^{n}\sum_{j=1}^{m}\sum_{k=1}^{r}(Y_{ijk} - \bar{Y})^2$$

$$\text{Between-Groups Sum of Squares (SSB):} \quad \text{SSB} = \sum_{i=1}^{n} m(Y_{i\cdot\cdot} - \bar{Y})^2$$

$$\text{Within-Groups Sum of Squares (SSW):} \quad \text{SSW} = \sum_{i=1}^{n}\sum_{j=1}^{m} r(Y_{ij\cdot} - Y_{i\cdot\cdot})^2$$

Example: Compare the mean scores of students across different teaching methods using ANOVA

ANOVA is a powerful statistical tool for comparing means across multiple groups, providing insights into group differences and interactions.

Chapter 9

Multivariate Statistics

9.1 Multivariate Normal Distribution

This section introduces key concepts related to the multivariate normal distribution in a quick and memorable way:

$$\text{Multivariate Normal PDF:} \quad f(\mathbf{x}; ^-, \boldsymbol{\Sigma})$$

$$= \frac{1}{(2\pi)^{p/2}|\boldsymbol{\Sigma}|^{1/2}} \exp\left(-\frac{1}{2}(\mathbf{x} - ^-)^T \boldsymbol{\Sigma}^{-1}(\mathbf{x} - ^-)\right)$$

$$\text{Mean Vector:} \quad ^- = \begin{bmatrix} \mu_1 \\ \mu_2 \\ \vdots \\ \mu_p \end{bmatrix}$$

$$\text{Covariance Matrix:} \quad \boldsymbol{\Sigma} = \begin{bmatrix} \sigma_{11} & \sigma_{12} & \cdots & \sigma_{1p} \\ \sigma_{21} & \sigma_{22} & \cdots & \sigma_{2p} \\ \vdots & \vdots & \ddots & \vdots \\ \sigma_{p1} & \sigma_{p2} & \cdots & \sigma_{pp} \end{bmatrix}$$

Example: Model the joint distribution of height, weight, and age of a population using a multivariate normal distribution

The multivariate normal distribution is a powerful tool for modeling joint distributions of multiple variables, providing insights into their dependencies and variability.

9.2   Principal Component Analysis (PCA)

This section introduces key concepts related to Principal Component Analysis in a quick and memorable way:

$$\text{Covariance Matrix:} \quad \mathbf{S} = \frac{1}{n-1}(\mathbf{X} - \bar{\mathbf{X}})^T(\mathbf{X} - \bar{\mathbf{X}})$$

$$\text{Eigenvalue Decomposition:} \quad \mathbf{S} = \mathbf{V}\mathbf{\Lambda}\mathbf{V}^T$$

$$\text{Principal Components (PCs):} \quad \mathbf{Z} = \mathbf{X}\mathbf{V}$$

$$\text{Proportion of Variance Explained:} \quad \text{Var}(\text{PC}_i) = \frac{\lambda_i}{\sum_{j=1}^{p}\lambda_j}$$

Example: Reduce the dimensionality of a dataset representing facial features using PCA

PCA is a dimensionality reduction technique that helps capture the most important features in a dataset, facilitating efficient analysis and visualization.

9.3   Canonical Correlation Analysis (CCA)

This section introduces key concepts related to Canonical Correlation Analysis in a quick and memorable way:

$$\text{Canonical Variates:} \quad \mathbf{U}_i = a_{i1}X_1 + a_{i2}X_2 + \cdots + a_{ip}X_p$$

$$\text{Canonical Correlation Coefficients:} \quad \rho_i = \sqrt{\frac{\lambda_i}{\lambda_i + 1}}$$

$$\text{Canonical Loadings:} \quad \mathbf{a}_i = \begin{bmatrix} a_{i1} \\ a_{i2} \\ \vdots \\ a_{ip} \end{bmatrix}$$

Example: Analyze the relationship between students' academic performance and their extracurricular activities using CCA.

Canonical Correlation Analysis helps identify the most correlated linear combinations of variables across two datasets, providing insights into their relationships.

9.4 Discriminant Analysis

This section introduces key concepts related to Discriminant Analysis in a quick and memorable way:

$$\text{Fisher's Linear Discriminant:} \quad \text{LD}(\mathbf{x}) = \frac{\mathbf{w}^T \mathbf{x}}{\|\mathbf{w}\|}$$

$$\text{Discriminant Function:} \quad \text{DF}_k(\mathbf{x}) = \mathbf{w}_k^T \mathbf{x} + w_{k0}$$

$$\text{Group Means:} \quad \mathbf{m}_k = \frac{1}{n_k} \sum_{i=1}^{n_k} \mathbf{x}_{ik}$$

Example: Classify species based on various morphological features using Discriminant Analysis

Discriminant Analysis aims to find the linear combinations of variables that best discriminate between different groups, facilitating classification tasks.

9.5 Cluster Analysis

This section introduces key concepts related to Cluster Analysis in a quick and memorable way:

$$\text{Distance Metric:} \quad d(\mathbf{X}_i, \mathbf{X}_j) = \sqrt{\sum_{k=1}^{p} (X_{ik} - X_{jk})^2}$$

Linkage Methods: Single, Complete, Average, Ward's method, etc.

Dendrogram: Visual representation of hierarchical clustering

Example: Group customers based on their purchasing behavior using Cluster Analysis

Cluster Analysis groups similar observations into clusters, aiding in the discovery of patterns and structures within datasets.

9.6   Multivariate Analysis of Variance (MANOVA)

This section introduces key concepts related to Multivariate Analysis of Variance in a quick and memorable way:

$$\text{Multivariate Response Vector:} \quad \mathbf{Y} = \begin{bmatrix} Y_1 \\ Y_2 \\ \vdots \\ Y_m \end{bmatrix}$$

$$\text{MANOVA Model:} \quad \mathbf{Y} = \mathbf{XB} + \mathbf{E}$$

$$\text{Wilks' Lambda:} \quad \Lambda = \frac{\det(\mathbf{E})}{\det(\mathbf{E} + \mathbf{XB})}$$

Example: Compare the effects of different treatments on multiple dependent variables, such as blood pressure, cholesterol, and heart rate

MANOVA extends ANOVA to handle multiple dependent variables simultaneously, providing a comprehensive analysis of group differences.

Chapter 10

Statistical Software

10.1 R Programming

This section introduces key concepts related to R Programming in a quick and memorable way:

Install a Package: install.packages("package_name")

Load a Package: library("package_name")

Create a Vector: vec ¡- c(1, 2, 3, 4, 5)

Perform a Linear Regression: lm_model ¡- lm(y x, data=df)

Generate a Plot: plot(x, y, main="Scatter Plot")

Example: Calculate the mean and standard deviation of a dataset using R

R is a versatile programming language for statistical computing and data analysis. It provides a wide range of functions and packages to facilitate exploratory data analysis and statistical modeling.

10.2   Python for Statistics

This section introduces key concepts related to using Python for Statistics in a quick and memorable way:

Import Libraries: import numpy as np, pandas as pd, matplotlib.pyplot as plt, seaborn as sns

Create a NumPy Array: data = np.array([1, 2, 3, 4, 5])

Perform Descriptive Statistics: mean = np.mean(data), std = np.std(data)

Create a DataFrame: df = pd.DataFrame("column_name": data)

Plot a Histogram: sns.histplot(df["column_name"]), plt.title("Histogram")

Example: Perform a t-test to compare the means of two groups in a dataset using Python

Python, with libraries like NumPy, pandas, and matplotlib, provides a powerful and accessible environment for statistical analysis and visualization.

10.3   SPSS Basics

This section introduces key concepts related to SPSS (Statistical Package for the Social Sciences) in a quick and memorable way:

Load Data: DATA LIST FILE="datafile.dat" /VAR x1 x2 x3.

Descriptive Statistics:
DESCRIPTIVES VARIABLES=x1 x2 x3 /STATISTICS=MEAN STDDEV MIN MAX.

Independent Samples t-test:
T-TEST GROUPS=group_variable /VARIABLES=dependent_variable.

Linear Regression:
REGRESSION /MISSING LISTWISE /STATISTICS COEFF OUTS R ANOVA /CRITERIA=PIN(.05) POUT(.10) /NOORIGIN /DEPENDENT dependent_variable /METHOD=ENTER independent_variable.

Graphical Representation:
GRAPH /SCATTERPLOT(BIVAR)=dependent_variable WITH independent_variable.

Example: Conduct a chi-square test to analyze the association
between two categorical variables in SPSS.

SPSS simplifies statistical analysis with an easy-to-use interface, making it
a popular tool for researchers and analysts in various fields.

10.4   SAS Fundamentals

This section introduces key concepts related to SAS (Statistical Analysis System)
in a quick and memorable way:

Data Step: DATA dataset_name;

PROC Step: PROC procedure_name DATA=dataset_name;

Variable Assignment: variable = expression;

Data Filtering: IF condition_expression THEN OUTPUT;

PROC MEANS: PROC MEANS DATA=dataset_name VAR=variable_list;

Example: Calculate the mean and standard deviation of a variable using SAS.

SAS provides a powerful environment for data manipulation, statistical analysis,
and reporting, making it widely used in various industries.

10.5   Excel for Data Analysis

This section introduces key concepts related to Excel for Data Analysis in a
quick and memorable way:

Sum Function: =SUM(range)

Average Function: =AVERAGE(range)

Standard Deviation Function: =STDEV(range)

Filtering Data: Use the Filter feature to select specific rows based on criteria.

PivotTables: Create PivotTables to summarize and analyze data easily.

Example: Calculate the total sales and average price per unit using Excel.

Excel offers a user-friendly interface for data analysis, allowing users to perform various calculations and visualizations efficiently.

10.6   Comparative Analysis of Statistical Software

This section provides a quick overview and comparison of key features among popular statistical software:

R: Extensive statistical libraries, open-source, and strong community support.

Python: Versatile programming language with powerful libraries like NumPy and pandas for data analysis.

SPSS: User-friendly interface, widely used in social sciences, and comprehensive statistical tests.

SAS: Robust for data manipulation, widely used in industries, and supports large datasets.

Excel: Accessible for basic data analysis, suitable for small to medium-sized datasets.

Example: Use R for advanced statistical modeling, Python for general-purpose data analysis, SPSS for social sciences research, SAS for large-scale industrial data, and Excel for simple calculations.

Choosing the right statistical software depends on factors such as the nature of the analysis, ease of use, and specific requirements.

Chapter 11

Quality Control and Six Sigma

11.1 Overview of Quality Control

This section provides a quick overview of key concepts in Quality Control:

$$\text{Population Mean } (\mu): \quad \mu = \frac{\sum_{i=1}^{N} X_i}{N}$$

$$\text{Sample Standard Deviation } (s): \quad s = \sqrt{\frac{\sum_{i=1}^{n}(X_i - \bar{X})^2}{n-1}}$$

Control Chart: A graphical tool to monitor and maintain the stability of a process over time.

$$\text{Process Capability } (\text{Cp}): \quad \text{Cp} = \frac{\text{USL} - \text{LSL}}{6s}$$

Statistical Process Control (SPC):   A set of statistical methods to monitor and control a process.

Example: Use control charts to monitor the weights of packaged products and ensure they meet specifications.

Quality Control ensures that processes meet desired standards and helps organizations maintain consistency and efficiency.

11.2   Statistical Process Control (SPC)

Control Chart Formulas:

$$X\text{-Bar Chart:} \quad \bar{X} = \frac{\sum_{i=1}^{n} X_i}{n}, \quad \text{CL}$$
$$= \bar{X}, \quad \text{UCL}$$
$$= \bar{X} + A_2 \times \frac{\sigma}{\sqrt{n}}, \quad \text{LCL}$$
$$= \bar{X} - A_2 \times \frac{\sigma}{\sqrt{n}}$$

$$\text{R Chart:} \quad R = \max(X_i) - \min(X_i), \quad \text{CL}$$
$$= R, \quad \text{UCL}$$
$$= D_4 \times R, \quad \text{LCL}$$
$$= D_3 \times R$$

Process Capability Indices:

$$\text{Cp:} \quad Cp = \frac{\text{USL} - \text{LSL}}{6\sigma}$$

$$\text{Cpk:} \quad Cpk = \min\left(\frac{\text{USL} - \bar{X}}{3\sigma}, \frac{\bar{X} - \text{LSL}}{3\sigma}\right)$$

Example:

Use an X-Bar Chart to monitor the mean and an R Chart to monitor the range of a manufacturing process.

Statistical Process Control involves using control charts and capability indices to monitor and improve the stability and capability of a process.

11.3   Control Charts

Control Charts are essential tools in Statistical Process Control (SPC) for monitoring and maintaining the stability of a process. Key formulas include:

X-Bar Chart:

$$\bar{X} = \frac{\sum_{i=1}^{n} X_i}{n}$$

$$CL = \bar{X}$$

$$UCL = \bar{X} + A_2 \times \frac{\sigma}{\sqrt{n}}$$

$$LCL = \bar{X} - A_2 \times \frac{\sigma}{\sqrt{n}}$$

R Chart:

$$R = \max(X_i) - \min(X_i)$$

$$CL = R$$

$$UCL = D_4 \times R$$

$$LCL = D_3 \times R$$

Here, $\bar{X}$ represents the sample mean, σ is the population standard deviation, n is the sample size, R is the range, and A_2, D_3, and D_4 are constants.

Control Charts provide a visual representation of how a process is performing over time, aiding in the detection of trends, shifts, or unusual patterns.

11.4 Capability Indices

Capability Indices assess how well a process meets specified requirements. Key formulas include:

Cp:

$$Cp = \frac{USL - LSL}{6\sigma}$$

Cpk:

$$Cpk = \min\left(\frac{USL - \bar{X}}{3\sigma}, \frac{\bar{X} - LSL}{3\sigma}\right)$$

Here, $\bar{X}$ represents the sample mean, σ is the population standard deviation, and USL and LSL are the upper and lower specification limits.

Cp measures the overall process capability, while Cpk considers both the mean and variability.

11.5 Six Sigma Methodology

Six Sigma is a data-driven methodology for process improvement. Key components include:

Defects per Million Opportunities (DPMO):

$$DPMO = \frac{\text{Number of Defects}}{\text{Total Opportunities}} \times 1,000,000$$

Sigma Level (Z):

$$Z = \frac{\text{USL or LSL} - \bar{X}}{\sigma}$$

Process Sigma (σ):

$$\sigma = \frac{\text{Range}}{d_2}$$

Here, $\bar{X}$ is the process mean, σ is the process standard deviation, and USL/LSL are upper/lower specification limits.

Six Sigma aims for a DPMO of less than 3.4, corresponding to a Sigma Level greater than 6.

11.6 Lean and Six Sigma Integration

The integration of Lean and Six Sigma combines efficiency and effectiveness. Key concepts include:

Lead Time Reduction:

$$\text{Lead Time} = \frac{\text{Work in Process}}{\text{Throughput Rate}}$$

Process Cycle Efficiency (PCE):

$$PCE = \frac{\text{Value-Added Time}}{\text{Total Lead Time}} \times 100\%$$

Overall Equipment Efficiency (OEE):

$$OEE = \text{Availability} \times \text{Performance} \times \text{Quality}$$

Here, the lead time is the time it takes to complete a process, and PCE and OEE assess efficiency and equipment utilization.

Lean focuses on waste reduction, while Six Sigma targets process variation. Integrating both enhances overall process performance.

Chapter 12

Survival Analysis

12.1 Survival Functions

Survival Functions are used in survival analysis to model time until an event occurs. Key components include:

Survival Probability (S(t)):

$$S(t) = P(T > t)$$

Hazard Function ($\lambda(t)$):

$$\lambda(t) = \lim_{\delta t \to 0} \frac{P(t \leq T < t + \delta t \mid T \geq t)}{\delta t}$$

Cumulative Hazard Function ($\Lambda(t)$):

$$\Lambda(t) = \int_0^t \lambda(u) \, du$$

Here, T is the random variable representing time until an event, and t is a specific time point.

Survival analysis is commonly used in medical research to analyze time until a patient experiences an event.

12.2 Hazard Functions

Hazard Functions describe the instantaneous failure rate in survival analysis. Key components include:

Hazard Function ($\lambda(t)$):

$$\lambda(t) = \lim_{\delta t \to 0} \frac{P(t \leq T < t + \delta t \mid T \geq t)}{\delta t}$$

Cumulative Hazard Function ($\Lambda(t)$):

$$\Lambda(t) = \int_0^t \lambda(u) \, du$$

Survival Probability (S(t)):

$$S(t) = \exp\left(-\Lambda(t)\right)$$

Here, T is the random variable representing time until an event, and t is a specific time point.

Hazard Functions help quantify the risk of an event occurring at a specific time in survival analysis.

12.3 Kaplan-Meier Estimator

The Kaplan-Meier Estimator is a non-parametric method for estimating survival functions from censored data. Key components include:

Survival Probability (Kaplan-Meier Estimate):

$$\hat{S}(t) = \prod_{i:t_i \leq t} \left(1 - \frac{d_i}{n_i}\right)$$

Hazard Function (Complementary Log-Log):

$$\lambda(t) = -\ln\left(1 - \frac{d_i}{n_i}\right)$$

Here, t is the time, t_i are event times, d_i is the number of events at time t_i, and n_i is the number of individuals at risk just before t_i.

The Kaplan-Meier Estimator is widely used in survival analysis to handle right-censored data.

12.4 Cox Proportional Hazards Model

The Cox Proportional Hazards Model is a semi-parametric model for survival analysis. Key components include:

Hazard Function (Cox Model):

$$\lambda(t, X) = \lambda_0(t) \cdot \exp(\beta_1 X_1 + \beta_2 X_2 + \ldots + \beta_p X_p)$$

Survival Probability (Cox Model):

$$\hat{S}(t, X) = \exp\left(-\int_0^t \lambda_0(u) \cdot \exp(\beta_1 X_1 + \beta_2 X_2 + \ldots + \beta_p X_p)\, du\right)$$

Here, $\lambda_0(t)$ is the baseline hazard function, X represents covariates, and $\beta_1, \beta_2, \ldots, \beta_p$ are coefficients.

The Cox Proportional Hazards Model allows for the assessment of the impact of covariates on survival probabilities.

12.5 Parametric Survival Models

Parametric Survival Models use specific distribution assumptions for survival times. Key models include:

Exponential Model:

$$\lambda(t) = \frac{1}{\theta} \quad \text{(Constant Hazard)}$$

Weibull Model:

$$\lambda(t) = \frac{\alpha}{\theta}\left(\frac{t}{\theta}\right)^{\alpha-1} \quad \text{(Generalization of Exponential)}$$

Log-Normal Model:

$$\log(T) \sim \mathcal{N}(\mu, \sigma^2) \quad \text{(Logarithm of Survival Time is Normally Distributed)}$$

Here, $\lambda(t)$ is the hazard function, θ is a scale parameter, α is the shape parameter, and μ, σ^2 are the mean and variance of the log-transformed survival time.

Parametric models provide a flexible approach to modeling survival data under specific distributional assumptions.

12.6   Time-Dependent Covariates

Time-Dependent Covariates in survival analysis allow covariate values to change over time. Key considerations include:

Model with Time-Dependent Covariates:

$$\lambda(t, X(t)) = \lambda_0(t) \cdot \exp(\beta_1 X_1(t) + \beta_2 X_2(t) + \ldots + \beta_p X_p(t))$$

Survival Probability with Time-Dependent Covariates:

$$\hat{S}(t, X(t)) = \exp\left(-\int_0^t \lambda_0(u) \cdot \exp(\beta_1 X_1(u) + \beta_2 X_2(u) + \ldots + \beta_p X_p(u)) \, du\right)$$

Here, $X(t)$ represents time-dependent covariates, and $X_1(t), X_2(t), \ldots, X_p(t)$ are the individual covariate functions.

Time-Dependent Covariates account for changes in covariate values during the course of the study in survival models.

Chapter 13

Meta-Analysis

13.1 Introduction to Meta-Analysis

Meta-Analysis combines results from multiple studies to draw overall conclusions. Key components include:

Weighted Average Effect Size:

$$\hat{\theta} = \frac{\sum_{i=1}^{k} w_i \cdot \theta_i}{\sum_{i=1}^{k} w_i}$$

Variance of the Combined Effect:

$$\mathrm{Var}(\hat{\theta}) = \frac{1}{\sum_{i=1}^{k} w_i}$$

Here, θ_i is the effect size from study i, and w_i is the weight assigned to study i based on its precision.

Meta-Analysis provides a quantitative synthesis of evidence across studies, yielding a more robust estimate of the overall effect.

13.2 Fixed and Random Effects Models

Fixed and Random Effects Models are used in Meta-Analysis to account for heterogeneity among studies. Key formulations include:

Fixed Effects Model:

$$\hat{\theta}_{\text{FE}} = \frac{\sum_{i=1}^{k} w_i \cdot \theta_i}{\sum_{i=1}^{k} w_i}$$

Variance of the Combined Effect (Fixed Effects):

$$\text{Var}(\hat{\theta}_{\text{FE}}) = \frac{1}{\sum_{i=1}^{k} w_i}$$

Random Effects Model:

$$\hat{\theta}_{\text{RE}} = \frac{\sum_{i=1}^{k} w_i \cdot \theta_i}{\sum_{i=1}^{k} w_i}$$

Variance of the Combined Effect (Random Effects):

$$\text{Var}(\hat{\theta}_{\text{RE}}) = \frac{\tau^2}{\sum_{i=1}^{k} w_i} + \frac{1 - \frac{\sum_{i=1}^{k} w_i}{\sum_{i=1}^{k} w_i^2}}{\sum_{i=1}^{k} w_i} \cdot \text{Var}(\theta_i)$$

Here, θ_i is the effect size from study i, w_i is the weight assigned to study i, and τ^2 represents the between-study variance.

Fixed Effects assume homogeneity, while Random Effects accommodate both within-study and between-study variability.

13.3   Publication Bias

Publication Bias occurs when studies with statistically significant results are more likely to be published. Key indicators include:

Funnel Plot Asymmetry:

$$\text{SE}(\hat{\theta}) = \sqrt{\frac{1}{\sum_{i=1}^{k} w_i}}$$

Egger's Test for Asymmetry:

$$\text{Egger's Statistic} = \frac{\sum_{i=1}^{k} \frac{\hat{\theta}_i}{\text{SE}(\hat{\theta}_i)}}{\sum_{i=1}^{k} \frac{1}{\text{SE}(\hat{\theta}_i)^2}}$$

Here, $\hat{\theta}_i$ is the effect size from study i, $\text{SE}(\hat{\theta}_i)$ is its standard error, and w_i is the weight assigned to study i.

Funnel plots and Egger's Test help assess the presence of publication bias in meta-analytic results.

13.4 Heterogeneity and Subgroup Analysis

Heterogeneity in Meta-Analysis refers to variability in effect sizes among studies. Key measures and methods include:

Cochran's Q Statistic:

$$Q = \sum_{i=1}^{k} w_i (\hat{\theta}_i - \hat{\theta}_{\text{overall}})^2$$

I^2 Statistic:

$$I^2 = \frac{Q - (k-1)}{Q} \times 100\%$$

DerSimonian and Laird Random Effects Model:

$$\hat{\theta}_{\text{RE}} = \frac{\sum_{i=1}^{k} w_i \cdot \hat{\theta}_i}{\sum_{i=1}^{k} w_i}$$

Subgroup Analysis:

$$\hat{\theta}_{\text{subgroup}} = \frac{\sum_{i=1}^{k} \sum_{j=1}^{n_i} w_{ij} \cdot \theta_{ij}}{\sum_{i=1}^{k} \sum_{j=1}^{n_i} w_{ij}}$$

Here, $\hat{\theta}_i$ is the effect size from study i, w_i is the weight assigned to study i, and w_{ij} are weights for subgroup analysis.

Cochran's Q, I^2, and subgroup analysis help explore and understand sources of heterogeneity in meta-analysis.

13.5 Meta-Regression

Meta-Regression extends meta-analysis to explore relationships between study characteristics and effect sizes. Key formulations include:

Meta-Regression Model:

$$\hat{\theta}_i = \beta_0 + \beta_1 X_{i1} + \beta_2 X_{i2} + \ldots + \beta_p X_{ip} + \epsilon_i$$

Weighted Meta-Regression Model:

$$\hat{\theta}_i = \beta_0 + \beta_1 X_{i1} + \beta_2 X_{i2} + \ldots + \beta_p X_{ip} + \epsilon_i$$

$$\text{Var}(\hat{\theta}_i) = \frac{1}{w_i}$$

Here, $\hat{\theta}_i$ is the effect size from study i, X_{ij} are study characteristics, $\beta_0, \beta_1, \ldots, \beta_p$ are regression coefficients, ϵ_i is the random error, and w_i is the weight assigned to study i.

Meta-Regression allows for the examination of how study characteristics impact the variability in effect sizes across studies.

13.6   Practical Considerations in Meta-Analysis

Meta-Analysis involves practical considerations to ensure robust results. Key aspects include:

Weighted Average Effect Size:

$$\hat{\theta}_{\text{overall}} = \frac{\sum_{i=1}^{k} w_i \cdot \hat{\theta}_i}{\sum_{i=1}^{k} w_i}$$

Variance of the Combined Effect:

$$\text{Var}(\hat{\theta}_{\text{overall}}) = \frac{1}{\sum_{i=1}^{k} w_i}$$

Confidence Intervals for Combined Effect:

$$\text{CI}(\hat{\theta}_{\text{overall}}) = \hat{\theta}_{\text{overall}} \pm z \times \sqrt{\text{Var}(\hat{\theta}_{\text{overall}})}$$

Here, $\hat{\theta}_i$ is the effect size from study i, w_i is the weight assigned to study i, z is the critical value for the desired confidence level.

Practical considerations include choosing appropriate effect size metrics, addressing heterogeneity, and assessing publication bias.

Chapter 14

Bayesian Statistics

14.1 Bayesian Inference Basics

Bayesian Inference provides a framework for updating beliefs based on new evidence. Key concepts include:

Bayes' Theorem:

$$P(\theta|X) = \frac{P(X|\theta) \cdot P(\theta)}{P(X)}$$

Posterior Distribution:

$$P(\theta|X) \propto P(X|\theta) \cdot P(\theta)$$

Prior Distribution:

$$P(\theta)$$

Likelihood Function:

$$P(X|\theta)$$

Here, θ represents parameters of interest, X is the observed data, and $P(\theta|X)$ is the posterior distribution.

Bayesian Inference combines prior knowledge (prior distribution) with new evidence (likelihood) to update beliefs (posterior distribution).

14.2   Prior and Posterior Distributions

In Bayesian Inference, understanding prior and posterior distributions is crucial. Key formulations include:

Bayes' Theorem:

$$P(\theta|X) = \frac{P(X|\theta) \cdot P(\theta)}{P(X)}$$

Posterior Distribution:

$$P(\theta|X) \propto P(X|\theta) \cdot P(\theta)$$

Prior Distribution:

$$P(\theta)$$

Likelihood Function:

$$P(X|\theta)$$

Here, θ represents parameters of interest, X is the observed data, and $P(\theta|X)$ is the posterior distribution.

Bayesian Inference involves updating prior beliefs (prior distribution) with new evidence (likelihood) to obtain updated beliefs (posterior distribution).

14.3   Markov Chain Monte Carlo (MCMC)

Markov Chain Monte Carlo is a method for sampling from complex probability distributions. Key elements include:

Metropolis-Hastings Algorithm:

$$P(\theta^{(t+1)}|\theta^{(t)}) = \min\left(1, \frac{P(\theta^{(t+1)}) \cdot P(X|\theta^{(t+1)})}{P(\theta^{(t)}) \cdot P(X|\theta^{(t)})}\right)$$

Gibbs Sampling:

$$P(\theta_j^{(t+1)}|\theta_1^{(t+1)}, \ldots, \theta_{j-1}^{(t+1)}, \theta_{j+1}^{(t)}, \ldots, \theta_p^{(t)})$$

Acceptance Probability:

$$P(\text{Accept}) = \min\left(1, \frac{P(\theta^{(t+1)})}{P(\theta^{(t)})}\right)$$

Here, $\theta^{(t)}$ represents the parameter values at iteration t, $P(\theta)$ is the target distribution, and $P(X|\theta)$ is the likelihood function.

MCMC methods, like Metropolis-Hastings and Gibbs Sampling, are valuable for approximating complex posterior distributions.

14.4 Bayesian Model Comparison

Bayesian Model Comparison involves comparing the evidence for different models. Key concepts include:

Bayes Factor:

$$\text{Bayes Factor} = \frac{P(X|M_1)}{P(X|M_2)}$$

Model Odds:

$$\text{Odds}_{M_1:M_2} = \frac{P(M_1|X)}{P(M_2|X)}$$

Posterior Model Probability:

$$P(M_i|X) = \frac{P(X|M_i) \cdot P(M_i)}{\sum_{j=1}^{k} P(X|M_j) \cdot P(M_j)}$$

Here, M_i represents the ith model, X is the observed data, $P(X|M_i)$ is the model evidence, and $P(M_i)$ is the prior model probability.

Bayesian Model Comparison helps in selecting the most plausible model based on the observed data.

14.5 Hierarchical Bayesian Models

Hierarchical Bayesian Models enable modeling complex structures with varying levels of uncertainty. Key formulations include:

Individual Level:

$$P(\theta_{ij}|X_{ij}) \propto P(X_{ij}|\theta_{ij}) \cdot P(\theta_{ij})$$

Group Level:

$$P(\mu_j|\alpha, \beta) \propto P(\mu_j|\alpha, \beta) \cdot P(\alpha, \beta)$$

Overall Structure:

$$P(\theta_{ij}, \mu_j, \alpha, \beta | X_{ij}) \propto P(X_{ij}|\theta_{ij}) \cdot P(\theta_{ij}) \cdot P(\mu_j|\alpha, \beta) \cdot P(\alpha, \beta)$$

Here, θ_{ij} represents individual parameters, μ_j represents group-level parameters, and X_{ij} is the observed data.

Hierarchical Bayesian Models allow for the incorporation of both individual and group-level uncertainties in the modeling process.

14.6   Bayesian Decision Theory

Bayesian Decision Theory provides a framework for making decisions under uncertainty. Key components include:

Decision Rule:

$$d^* = \arg\max_d \sum_\theta L(\theta, d) \cdot P(\theta|X)$$

Expected Loss:

$$\text{Expected Loss} = \sum_\theta L(\theta, d^*) \cdot P(\theta|X)$$

Risk Function:

$$R(d) = \sum_\theta L(\theta, d) \cdot P(\theta|X)$$

Here, d^* is the optimal decision, $L(\theta, d)$ is the loss function, $P(\theta|X)$ is the posterior distribution, and $R(d)$ is the risk associated with decision d.

Bayesian Decision Theory aims to minimize the expected loss or risk associated with decision-making.

Chapter 15

Statistical Ethics

15.1 Ethical Considerations in Data Collection

Ethical considerations in data collection are crucial for responsible research practices. Key principles include:

Informed Consent:

$$P(\text{Informed Consent}) = \frac{\text{Number of Participants Providing Informed Consent}}{\text{Total Number of Participants}}$$

Privacy Preservation:

$$P(\text{Privacy Preservation}) = \frac{\text{Number of Anonymized Data Instances}}{\text{Total Number of Data Instances}}$$

Data Security:

$$P(\text{Data Security}) = \frac{\text{Number of Securely Stored Data Sets}}{\text{Total Number of Data Sets}}$$

Here, $P(\text{Informed Consent})$, $P(\text{Privacy Preservation})$, and $P(\text{Data Security})$ represent the probabilities of adherence to ethical principles.

Ethical data collection ensures the protection of participants' rights and the responsible use of data.

15.2   Data Privacy and Confidentiality

Ensuring data privacy and confidentiality is essential in statistical analysis. Key considerations include:

K-anonymity:

$$\text{K-anonymity} = \min(\text{Number of Identifiers})$$

L-diversity:

$$\text{L-diversity} = \min(\text{Number of Sensitive Attribute Values})$$

T-closeness:

$$\text{T-closeness} = \max(\text{Distance Metric for Sensitivity})$$

Here, K-anonymity, L-diversity, and T-closeness are measures to protect individual privacy and ensure confidentiality in data.

Implementing these concepts helps in preventing re-identification and safeguarding sensitive information.

15.3   Publication Ethics

Publication ethics is vital for maintaining the integrity of scholarly work. Key principles include:

Plagiarism Detection:

$$\text{Plagiarism Detection} = \frac{\text{Number of Detected Plagiarized Instances}}{\text{Total Number of Instances}}$$

Authorship Integrity:

$$\text{Authorship Integrity} = \frac{\text{Number of Properly Acknowledged Authors}}{\text{Total Number of Authors}}$$

Reviewer Neutrality:

$$\text{Reviewer Neutrality} = \frac{\text{Number of Unbiased Reviewer Assessments}}{\text{Total Number of Reviews}}$$

Here, Plagiarism Detection, Authorship Integrity, and Reviewer Neutrality are metrics to ensure ethical publication practices.

Adhering to these principles promotes trust in the scientific community and upholds the quality of published research.

15.4 Reproducibility and Transparency

Ensuring reproducibility and transparency in research is crucial for scientific rigor. Key aspects include:

Reproducibility Rate:

$$\text{Reproducibility Rate} = \frac{\text{Number of Replicable Studies}}{\text{Total Number of Studies}}$$

Transparency Index:

$$\text{Transparency Index} = \frac{\text{Number of Transparently Reported Methods}}{\text{Total Number of Methods}}$$

Open Data Access:

$$\text{Open Data Access} = \frac{\text{Number of Accessible Datasets}}{\text{Total Number of Datasets}}$$

Here, Reproducibility Rate, Transparency Index, and Open Data Access are measures to enhance the reliability and openness of research.

Adopting these practices fosters trust in scientific findings and promotes the advancement of knowledge.

15.5 Responsible Conduct of Research

The responsible conduct of research involves ethical and professional behavior. Key principles include:

Integrity Index:

$$\text{Integrity Index} = \frac{\text{Number of Research Misconduct Cases}}{\text{Total Number of Research Projects}}$$

Collaboration Quotient:

$$\text{Collaboration Quotient} = \frac{\text{Number of Collaborative Research Initiatives}}{\text{Total Number of Research Initiatives}}$$

Adherence to Protocols:

$$\text{Adherence to Protocols} = \frac{\text{Number of Projects Following Ethical Protocols}}{\text{Total Number of Projects}}$$

Here, Integrity Index, Collaboration Quotient, and Adherence to Protocols are metrics to gauge responsible research conduct.

Following these principles ensures the reliability of research outcomes and contributes to the advancement of knowledge responsibly.

15.6   Case Studies in Statistical Ethics

Examining case studies in statistical ethics provides insights into real-world ethical dilemmas. Key considerations include:

Ethical Analysis:

$$\text{Ethical Analysis} = \frac{\text{Number of Cases with Ethical Resolutions}}{\text{Total Number of Cases}}$$

Informed Decision-making:

$$\text{Informed Decision-making} = \frac{\text{Number of Cases with Well-Informed Decisions}}{\text{Total Number of Cases}}$$

Ethical Impact Assessment:

$$\text{Ethical Impact Assessment} = \frac{\text{Number of Cases with Positive Ethical Impact}}{\text{Total Number of Cases}}$$

Here, Ethical Analysis, Informed Decision-making, and Ethical Impact Assessment are measures to evaluate ethical considerations in statistical practice.

Reflecting on these case studies enhances ethical awareness and guides practitioners in making sound ethical choices.

Chapter 16

Spatial Statistics

16.1 Spatial Autocorrelation

Spatial autocorrelation measures the similarity of values in neighboring locations. Key concepts include:

Moran's I Index:

$$I = \frac{n}{W} \frac{\sum_{i=1}^{n} \sum_{j=1}^{n} w_{ij}(x_i - \bar{x})(x_j - \bar{x})}{\sum_{i=1}^{n}(x_i - \bar{x})^2}$$

Geary's C Index:

$$C = \frac{(n-1)\sum_{i=1}^{n} \sum_{j=1}^{n} w_{ij}(x_i - x_j)^2}{2\sum_{i=1}^{n} \sum_{j=1}^{n} w_{ij}(x_i - \bar{x})^2}$$

Local Indicators of Spatial Association (LISA):

$$LISA_i = \frac{(x_i - \bar{x})}{S} \sum_{j=1}^{n} w_{ij}(x_j - \bar{x})$$

Here, I and C are global spatial autocorrelation indices, and $LISA_i$ represents the local spatial autocorrelation for each location i. w_{ij} denotes the spatial weight between locations i and j, n is the number of locations, x_i is the value at location i, and $\bar{x}$ is the mean value.

Understanding these indices helps assess spatial patterns and dependencies in datasets.

16.2   Geostatistics

Geostatistics is a branch of statistics that deals with spatial variability. Key concepts include:

Variogram:

$$\gamma(h) = \frac{1}{2N(h)} \sum_{i=1}^{N(h)} (z(x_i + h) - z(x_i))^2$$

Kriging:

$$\hat{z}(u) = \sum_{i=1}^{n} \lambda_i z(u_i)$$

Semi-Variogram:

$$\gamma(h) = \frac{1}{2N(h)} \sum_{i=1}^{N(h)} (z(x_i) - z(x_i + h))^2$$

Here, $\gamma(h)$ represents the variogram, $N(h)$ is the number of pairs of observations separated by distance h, $z(x_i)$ is the observed value at location x_i, and $\hat{z}(u)$ is the kriging estimate at an unsampled location u. The semi-variogram assesses the variance between observations at different distances.

Utilizing these geostatistical tools aids in understanding and predicting spatial patterns in data.

16.3   Point Pattern Analysis

Point pattern analysis is used to study the spatial distribution of points. Key concepts include:

Ripley's K-function:

$$K(t) = \frac{1}{n^2 \lambda} \sum_{i=1}^{n} \sum_{j=1}^{n} w_{ij}(t)$$

Pair Correlation Function:

$$g(r) = \frac{k(r)}{\lambda \pi r^2}$$

L Function:

$$L(t) = \sqrt{\frac{K(t)}{\pi}}$$

Here, n is the number of points, $w_{ij}(t)$ is a weight function, r is the distance between points, and λ is the spatial intensity. The K-function assesses the deviation of points from a random distribution, the pair correlation function evaluates clustering, and the L function is related to the K-function.

Using these functions helps characterize the spatial patterns and clustering of point data.

16.4 Spatial Regression Models

Spatial regression models are used to analyze relationships between variables while considering spatial dependencies. Key concepts include:

Spatial Lag Model:

$$y = \rho W y + X\beta + \varepsilon$$

Spatial Error Model:

$$y = X\beta + \lambda W \varepsilon$$

Spatial Durbin Model:

$$y = X\beta + \rho W y + \lambda W \varepsilon$$

Here, y is the dependent variable, X is the matrix of independent variables, β is the vector of coefficients, ε is the error term, and W is the spatial weight matrix. ρ and λ are parameters indicating spatial lag and spatial autocorrelation, respectively.

These models help account for spatial relationships when analyzing regression relationships in spatial data.

16.5   Cluster Detection

Cluster detection involves identifying spatial concentrations of events. Key methods include:

Kulldorff's Scan Statistic:

$$V = \max_C \frac{\text{Observed cases in } C - E[\text{Expected cases in } C]}{\sqrt{E[\text{Expected cases in } C]}}$$

Spatial Scan Statistic:

$$V = \max_C \frac{\text{Observed cases in } C}{E[\text{Expected cases in } C]}$$

Moran's I Spatial Autocorrelation Statistic:

$$I = \frac{n}{W} \frac{\sum_{i=1}^{n} \sum_{j=1}^{n} w_{ij}(x_i - \bar{x})(x_j - \bar{x})}{\sum_{i=1}^{n}(x_i - \bar{x})^2}$$

Here, V represents the test statistic, C denotes potential clusters, E is the expected value, n is the number of locations, w_{ij} is the spatial weight, and x_i is the observed value at location i.

These statistics aid in detecting clusters and spatial patterns in data.

16.6   Spatial Data Visualization

Spatial data visualization is crucial for understanding patterns and relationships. Key concepts include:

Choropleth Maps:

$$\text{Color}(i) = \frac{value_i - \min(values)}{\max(values) - \min(values)}$$

Heatmaps:

$$\text{Intensity}(i) = \frac{\text{count of points in region } i}{\text{max count of points in any region}}$$

Flow Maps:

$$\text{Flow Width}(i, j) = \frac{\text{flow intensity from location } i \text{ to location } j}{\text{max flow intensity}}$$

Here, $value_i$ represents the value at location i, $\min(values)$ and $\max(values)$ are the minimum and maximum values in the dataset, and count of points in region i is the number of data points in region i.

These visualization techniques help convey spatial information effectively.

Chapter 17

Machine Learning in
Statistics

17.1 Introduction to Machine Learning

Machine learning involves the development of algorithms that enable computers
to learn patterns from data. Key concepts include:

Linear Regression:

$$y = \beta_0 + \beta_1 x + \varepsilon$$

Logistic Regression:

$$P(Y = 1) = \frac{1}{1 + e^{-(\beta_0 + \beta_1 x)}}$$

Support Vector Machines (SVM):

$$\arg\min_{\mathbf{w}, b, \zeta} \frac{1}{2} \|\mathbf{w}\|^2 + C \sum_{i=1}^{n} \zeta_i$$

Here, y is the target variable, x is the input variable, ε is the error term,
$P(Y = 1)$ is the probability of class 1, $\mathbf{w}$ is the weight vector, b is the bias term,
and ζ represents slack variables in SVM.

These algorithms form the foundation of machine learning, addressing regression,
classification, and support vector machines.

17.2   Supervised Learning

Supervised learning involves training a model on labeled data to make predictions. Key algorithms include:

Linear Regression:

$$y = \beta_0 + \beta_1 x + \varepsilon$$

Logistic Regression:

$$P(Y = 1) = \frac{1}{1 + e^{-(\beta_0 + \beta_1 x)}}$$

Decision Trees:

$$\text{Splitting criterion: } J(D, v) = H(D) - \frac{|D_L|}{|D|} H(D_L) - \frac{|D_R|}{|D|} H(D_R)$$

Here, y is the target variable, x is the input variable, ε is the error term, $P(Y = 1)$ is the probability of class 1 in logistic regression, and $J(D, v)$ is the splitting criterion for decision trees.

These supervised learning algorithms are foundational for regression, classification, and decision-making tasks.

17.3   Unsupervised Learning

Unsupervised learning involves exploring data patterns without labeled outcomes. Key algorithms include:

K-Means Clustering:

$$J(C, \mu) = \sum_{i=1}^{k} \sum_{j=1}^{n_i} \|x_{ij} - \mu_i\|^2$$

Principal Component Analysis (PCA):

$$\text{Maximize: } \text{Var}(Y_1) = \frac{1}{n} \sum_{i=1}^{n} (\mathbf{y}_i^T \mathbf{v}_1)^2$$

Hierarchical Clustering:

$$\text{Linkage Criterion: } d(C_i, C_j) = \min_{x_i \in C_i, x_j \in C_j} \|x_i - x_j\|$$

Here, $J(C, \mu)$ is the K-Means objective function, $\text{Var}(Y_1)$ represents the variance of the first principal component in PCA, and $d(C_i, C_j)$ is the linkage criterion for hierarchical clustering.

These unsupervised learning techniques aid in identifying hidden structures and patterns within data.

17.4 Feature Selection and Engineering

Feature selection and engineering are vital in enhancing model performance. Key techniques include:

Information Gain (IG):

$$\text{IG}(D, A) = H(D) - \sum_{v \in \text{Values}(A)} \frac{|D_v|}{|D|} H(D_v)$$

Principal Component Analysis (PCA):

$$\text{Var}(Y_1) = \frac{1}{n} \sum_{i=1}^{n} (\mathbf{y}_i^T \mathbf{v}_1)^2$$

Polynomial Feature Engineering:

$$\text{Degree-2 Polynomial: } (a + b)^2 = a^2 + 2ab + b^2$$

Here, $\text{IG}(D, A)$ is the information gain for feature A, $\text{Var}(Y_1)$ represents the variance of the first principal component in PCA, and the polynomial feature engineering introduces quadratic terms.

These techniques contribute to selecting informative features and creating new ones to improve model performance.

17.5 Model Evaluation Metrics

Model performance is assessed using various metrics. Key evaluation measures include:

Mean Squared Error (MSE):

$$MSE = \frac{1}{n} \sum_{i=1}^{n} (y_i - \hat{y}_i)^2$$

Accuracy:

$$\text{Accuracy} = \frac{\text{Number of Correct Predictions}}{\text{Total Number of Predictions}}$$

Precision:

$$\text{Precision} = \frac{\text{True Positives}}{\text{True Positives} + \text{False Positives}}$$

Here, y_i is the true target value, $\hat{y}_i$ is the predicted value, and True Positives, False Positives, and False Negatives are components of the confusion matrix.

These metrics provide insights into the model's ability to make accurate predictions and its performance in classification tasks.

17.6   Applications of Machine Learning in Statistics

Machine learning techniques find various applications in statistics. Key areas include:

Predictive Modeling:

$$\text{Model: } y = f(x) + \varepsilon$$

Classification Tasks:

$$\text{Logistic Regression: } P(Y = 1) = \frac{1}{1 + e^{-(\beta_0 + \beta_1 x)}}$$

Clustering:

$$J(C, \mu) = \sum_{i=1}^{k} \sum_{j=1}^{n_i} \|x_{ij} - \mu_i\|^2$$

Here, y is the target variable, x is the input variable, ε is the error term, $P(Y = 1)$ is the probability of class 1 in logistic regression, and $J(C, \mu)$ is the K-Means objective function.

These applications showcase the versatility of machine learning in solving statistical problems.

Chapter 18

Statistical Network Analysis

18.1 Graph Theory Basics

Graph theory is fundamental in various fields. Key concepts include:

Graph Definition:

$$G = (V, E)$$

Degree of a Vertex:

$$\text{Degree}(v) = \sum_{u \in N(v)} 1$$

Adjacency Matrix:

$$A_{ij} = \begin{cases} 1 & \text{if } (v_i, v_j) \in E \\ 0 & \text{otherwise} \end{cases}$$

Here, G is a graph with vertices V and edges E, $\text{Degree}(v)$ is the degree of vertex v, and A_{ij} is the entry in the adjacency matrix.

These basics lay the foundation for understanding and analyzing complex networks using graph theory.

18.2   Centrality Measures

Centrality measures quantify the importance of nodes in a graph. Key measures include:

Degree Centrality:

$$C_D(v) = \frac{\text{Degree}(v)}{\text{Total Nodes} - 1}$$

Closeness Centrality:

$$C_C(v) = \frac{1}{\sum_{u \neq v} d(v, u)}$$

Betweenness Centrality:

$$C_B(v) = \sum_{s \neq v \neq t} \frac{\sigma_{st}(v)}{\sigma_{st}}$$

Here, $C_D(v)$ is the degree centrality, $C_C(v)$ is the closeness centrality, and $C_B(v)$ is the betweenness centrality.

These measures help identify nodes with significant influence or connectivity within a graph.

18.3   Community Detection

Community detection aims to identify groups of tightly connected nodes in a graph. Key methods include:

Modularity Function:

$$Q = \frac{1}{2m} \sum_{ij} \left(A_{ij} - \frac{k_i k_j}{2m} \right) \delta(c_i, c_j)$$

Louvain Method:

$$\Delta Q = \sum_i \left[\frac{\sum_{in} A_{in} \delta(c_i, c_n) - k_i^{in} \frac{\sum_{tot} A_{tot} \delta(c_i, c_{tot})}{2m}}{2m} \right]$$

Girvan-Newman Algorithm:

$$\text{Edge Betweenness}(e) = \sum_{s,t \in V} \frac{\sigma_{st}(e)}{\sigma_{st}}$$

Here, Q is the modularity, ΔQ is the change in modularity, and Edge Betweenness measures the centrality of edges.

These methods aid in uncovering meaningful structures within complex networks.

18.4 Network Models

Network models describe various types of connections and interactions. Key models include:

Erdős-Rényi Model:

$$P(A_{ij}) = p$$

Watts-Strogatz Model:

$$P(A_{ij}) = \begin{cases} p, & \text{with probability } \beta \\ 0, & \text{with probability } 1 - \beta \end{cases}$$

Barabási-Albert Model:

$$P(A_{ij}) \propto k_i^{\alpha}$$

Here, $P(A_{ij})$ represents the probability of an edge between nodes i and j, and k_i is the degree of node i.

These models capture different aspects of network structures and dynamics.

18.5 Temporal Network Analysis

Temporal network analysis deals with dynamic interactions over time. Key concepts include:

Temporal Paths:

$$P(u, v, t) = \{(u, t_1), \ldots, (v, t_k)\}$$

Temporal Centrality:

$$C_T(v) = \sum_{u \neq v} \frac{\text{Number of temporal paths passing through } v}{\text{Total number of temporal paths}}$$

Temporal Clustering Coefficient:

$$\text{TCC}(v) = \frac{\text{Number of closed temporal triplets at } v}{\text{Number of connected temporal triplets at } v}$$

Here, $P(u, v, t)$ represents a temporal path from node u to v at different time points, $C_T(v)$ is the temporal centrality, and TCC is the temporal clustering coefficient.

These measures capture temporal aspects of network dynamics.

18.6   Applications in Social Networks

Social network analysis finds various applications in understanding social structures. Key concepts include:

Homophily Index:

$$H = \frac{\sum_{i,j} A_{ij} \delta(x_i, x_j)}{\sum_{i,j} A_{ij}}$$

Centrality in Social Networks:

$$C(v) = \frac{\text{Number of shortest paths passing through } v}{\text{Total number of shortest paths}}$$

Triadic Closure:

$$T = \frac{\text{Number of closed triads}}{\text{Total number of triads}}$$

Here, H measures homophily, $C(v)$ is centrality, and T represents triadic closure.

These concepts help analyze social interactions and structural patterns within social networks.

Chapter 19

Environmental Statistics

19.1 Spatial and Temporal Environmental Data

Analyzing spatial and temporal environmental data involves key mathematical concepts:

Spatial Autocorrelation:

$$\text{Spatial Autocorrelation} = \frac{\sum_{i,j} w_{ij}(x_i - \bar{x})(x_j - \bar{x})}{\sum_i (x_i - \bar{x})^2}$$

Geostatistics:

$$Z(d) = \frac{\sum_{i=1}^{n(d)} w_i Z_i}{\sum_{i=1}^{n(d)} w_i}$$

Point Pattern Analysis:

$$K(d) = \frac{\text{Number of points within distance } d}{\text{Expected number of points within distance } d}$$

Here, w_{ij} represents spatial weights, $Z(d)$ is the semivariogram, and $K(d)$ is the spatial point pattern function.

These methods aid in understanding spatial and temporal patterns in environmental data.

19.2   Air Quality Modeling

Air quality modeling involves key mathematical expressions to estimate pollutant concentrations:

Gaussian Dispersion Model:

$$C(x, y, z) = \frac{Q}{2\pi\sigma_x\sigma_y\sigma_z} \exp\left(-\frac{1}{2}\left(\frac{(x - x_0)^2}{\sigma_x^2} + \frac{(y - y_0)^2}{\sigma_y^2} + \frac{(z - z_0)^2}{\sigma_z^2}\right)\right)$$

Box Model:

$$C(t) = \frac{E}{\tau}\left(1 - \exp\left(-\frac{t}{\tau}\right)\right)$$

Chemical Transport Models:

$$\frac{\partial C}{\partial t} + \nabla \cdot (\mathbf{u}C) = D\nabla^2 C + S$$

Here, $C(x, y, z)$ represents pollutant concentration, Q is the emission rate, and $\mathbf{u}$ is the wind vector.

These models help simulate and understand the dispersion of pollutants in the air.

19.3   Water Quality Analysis

Analyzing water quality involves key mathematical expressions to assess various parameters:

Water Quality Index (WQI):

$$WQI = \left(\sum_{i=1}^{n} w_i \frac{I_i}{I_{\max}}\right) \times 100$$

Dissolved Oxygen (DO) Saturation:

$$DO\% = \frac{\text{Measured DO}}{\text{DO Saturation}} \times 100$$

Biochemical Oxygen Demand (BOD):

$$BOD = \frac{\text{Initial DO} - \text{Final DO}}{\text{Volume of Water Sample}}$$

These formulas provide a quick assessment of water quality, with WQI summarizing multiple parameters.

19.4 Ecological Statistics

Ecological statistics involves key mathematical expressions to analyze ecological patterns:

Species Diversity Index (Shannon-Wiener Index):

$$H' = -\sum_{i=1}^{S} p_i \ln(p_i)$$

Species Richness:

$$S = \frac{N}{\hat{S}}$$

Simpson's Diversity Index:

$$D = \frac{\sum_{i=1}^{S} n_i(n_i - 1)}{N(N - 1)}$$

These formulas help quantify biodiversity and ecological patterns in a given environment.

19.5 Climate Change Statistics

Understanding climate change involves key mathematical expressions to analyze trends and impacts:

Temperature Anomaly:

$$T_a = T - \bar{T}$$

Trend Analysis (Linear Regression):

$$y = mx + b$$

Climate Extremes Index (CEI):

$$CEI = \frac{1}{N} \sum_{i=1}^{N} R_i$$

These formulas provide insights into temperature anomalies, trends, and extreme climate events.

19.6   Environmental Impact Assessment

Assessing environmental impact involves key mathematical expressions to quantify effects:

Environmental Impact Index (EII):

$$EII = \frac{\sum_{i=1}^{n}(W_i \times I_i)}{\sum_{i=1}^{n} W_i}$$

Risk Assessment:

$$Risk = \text{Probability} \times \text{Consequence}$$

Pollution Load Index (PLI):

$$PLI = \frac{\sum_{i=1}^{n} C_i}{n \times C_{\text{ref}}}$$

These formulas aid in quantifying environmental impact, assessing risks, and estimating pollution loads.

Chapter 20

Statistical Genetics

20.1 Basic Concepts in Genetics

Understanding genetics involves fundamental mathematical expressions to describe
key concepts:

Punnett Square for Mendelian Inheritance:

$$P(\text{Genotype}) = \frac{\text{Number of favorable outcomes}}{\text{Total possible outcomes}}$$

Hardy-Weinberg Equilibrium:

$$p^2 + 2pq + q^2 = 1$$

Genetic Diversity (Heterozygosity):

$$H = 1 - \sum_{i=1}^{n} p_i^2$$

These formulas provide a quick overview of genetic concepts, including inheritance
patterns and population genetics.

20.2 Linkage Analysis

Linkage analysis in genetics involves essential mathematical expressions to assess
genetic linkage:

Logarithm of Odds (LOD) Score:

$$\text{LOD} = \log_{10}\left(\frac{\text{Probability of Linkage}}{\text{Probability of No Linkage}}\right)$$

Recombination Frequency (θ):

$$\theta = \frac{\text{Number of recombinant gametes}}{\text{Total number of gametes}}$$

Two-Point Linkage Analysis:

$$\text{LOD}_{12} = \log_{10}\left(\frac{L(\theta = 0)}{L(\theta = 0.5)}\right)$$

These formulas provide a quick overview of mathematical concepts in linkage analysis.

20.3   Association Studies

Association studies in genetics involve crucial mathematical expressions to assess relationships between genetic variants and traits:

Odds Ratio (OR) for Case-Control Studies:

$$\text{OR} = \frac{(\text{Odds of Exposure in Cases})/(\text{Odds of Exposure in Controls})}{(\text{Odds of No Exposure in Cases})/(\text{Odds of No Exposure in Controls})}$$

Chi-Square Test for Allelic Association:

$$\chi^2 = \sum \frac{(O_i - E_i)^2}{E_i}$$

Linkage Disequilibrium (LD) Measures:

$$D = P_{11}P_{22} - P_{12}P_{21}$$

These formulas provide a quick overview of mathematical concepts in association studies.

20.4   Genome-Wide Association Studies (GWAS)

Genome-Wide Association Studies (GWAS) involve complex statistical methods. Let's briefly explore the key concepts:

Single Nucleotide Polymorphism (SNP) Association:

$$\text{Odds Ratio (OR)} = \frac{AD \cdot BC}{AC \cdot BD}$$

Allelic Association Test:

$$\chi^2 = \frac{(n_{11}n_{22} - n_{12}n_{21})^2 \cdot N}{(n_{11} + n_{12})(n_{21} + n_{22})(n_{11} + n_{21})(n_{12} + n_{22})}$$

Genomic Control:

$$\lambda_{GC} = \frac{\text{Median of Observed } \chi^2}{\text{Median of Expected } \chi^2}$$

These formulas provide a rapid understanding of the statistical methods employed in GWAS.

20.5   Genomic Data Analysis

Genomic data analysis involves various statistical techniques. Let's briefly explore some key concepts:

Genomic Variability:

$$\text{Genomic Variance} = \frac{\sum_{i=1}^{n}(x_i - \bar{x})^2}{n - 1}$$

Genomic Correlation:

$$\text{Genomic Correlation Coefficient} = \frac{\text{Cov}(X, Y)}{\sqrt{\text{Var}(X) \cdot \text{Var}(Y)}}$$

Genomic Regression:

$$Y = \beta_0 + \beta_1 X + \epsilon$$

Genomic Principal Component Analysis (PCA):

$$\text{PCA}(\mathbf{X}) = \mathbf{T} = \mathbf{XW}$$

These formulas provide a quick overview of the mathematical foundations for genomic data analysis.

20.6   Ethical and Privacy Considerations in Genetic Studies

Understanding the ethical and privacy aspects of genetic studies is crucial. Let's discuss key considerations:

Informed Consent: In genetic studies, participants should provide informed consent, denoted as C:

$$C = \frac{E}{I}$$

Privacy Preservation: Protecting individual privacy is essential, and measures such as anonymization can be represented as:

$$\text{Anonymized Data} = \text{Genetic Data} \times \text{Privacy Matrix}$$

Ethical Guidelines: Adhering to ethical guidelines ensures responsible conduct. The ethical index EI can be defined as:

$$EI = \frac{\text{Number of Ethical Principles Followed}}{\text{Total Ethical Principles}}$$

These equations offer a brief overview of the ethical and privacy considerations in genetic studies.